FIRE
PROTECTION & SUPPRESSION

Robert E. Colburn

Division Fire Marshal
Chicago Fire Department

Instructor, Fire Science Program
Triton Community College

McGraw-Hill Book Company

New York • St. Louis • Dallas • San Francisco • Auckland •Düsseldorf
Johannesburg • Kuala Lumpur • London • Mexico • Montreal • New Delhi
Panama • Paris • São Paulo • Singapore • Sydney • Tokyo • Toronto

Library of Congress Cataloging in Publication Data

Colburn, Robert E
   Fire protection & suppression.

  Includes index.
  1. Fire prevention. 2. Fire extinction. 3. Fire-
departments. I. Title.
TH9241.C64   628.9'2   74-30481
ISBN 0-07-011680-6

FIRE PROTECTION AND SUPPRESSION

1234567890 KPKP 78321098765

Cover Photo Courtesy of
  N.Y.F.D. Photo Unit
  Fr. F. Murphy

Title Page Photo Courtesy of
  N.Y.F.D. Photo Unit
  Fr. A. Delle Donne

The editors for this book were *Carole O'Keefe* and *Alice V. Manning,* the designer was
Tracy Glasner, and the production supervisor was *Phyllis D. Lemkowitz.* It was set in
Helvetica by John C. Meyer & Son, Inc.
Printed and bound by Kingsport Press, Inc.

CONTENTS

PREFACE

"No man is an island"—John Donne (1573–1631)

The fire protection service is deeply in debt to many agencies who have helped develop and advance fire protection measures. The list is long. Some are nonprofit organizations such as the National Fire Protection Association; others are insurance service organizations, fire testing and research laboratories, and certain government agencies. All have contributed much to fire safety and protection.

Firefighters themselves have begun to flex their muscles. They have been challenged by fire problems of modern technology and they are responding. They are acquiring the greater knowledge needed to combat the new and more sophisticated fire hazards. A striving toward professionalism is the goal of the new breed of firefighters.

This striving is emphasized in the many courses in fire science now being offered in community colleges. It is further emphasized by individual studying and by study groups working in fire departments beyond the reach of, or in conjunction with, community colleges.

The intent of this text is to present basic knowledge in the broad field of fire technology to the student who is participating in a structured or an unstructured study group. The book provides knowledge of the composition, causes, and behavior of fire; increases the reader's awareness of extinguishing agents and equipment used in the extinguishment of fire; and builds understanding of the fire department organization, its procedures, and the tactics it uses in the extinguishment of fire.

The student who reads no more than this text should have a good grasp of the firefighter's task and role. The student who progresses to greater in-depth studies in specific areas should have established a solid base from which to specialize.

The prospective reader has varying needs and interests. He may be the future entrant to the fire service, the present firefighter, or employed or seeking employment in a field allied with fire protection. To provide for these differences, discussion topics and research projects follow each chapter so that specific concentration topics can be tailored to the individual.

The order of chapter presentation is a consensus of the author's opinion and the polled opinion of a select group. The individual instructor or the self-study student may very well choose to vary chapter sequence, taking first those chapters of greatest importance to his or her particular course, need, or desire. Each chapter is capable of standing on its own and is not necessarily contingent upon another.

One basic text cannot attempt to encompass regional differences in terminology, methods of operation, and variations in codes and ordinances. Yet, underlying the text must be basic beliefs, descriptions of practices and procedures, guidelines, and recommendations appropriate to all in the fire service; errors or omissions are the responsibility of the author. Credit must be given to the allied interests and those in the fire service who have contributed the basic tenets and who have made constructive critiques of this text.

Special thanks to all the "C's," who with patience, tolerance, and love gave so much of themselves in support, encouragement, and time and, most importantly, gave true meaning to the word "family."

Robert E. Colburn

INTRODUCTION

THE GENESIS OF FIRE

The technology of the civilized world takes giant strides in the life span of a single individual. We are constantly on the threshold of new discoveries; using the technology of the present, the human brain opens closed doors to tomorrow's greater advances. The foundation stone at the base of all technological improvements that have brought man to the civilization of the twentieth century was the use and control of fire.

Greek mythology relates that the world was without fire until Prometheus resolved to improve mankind's earthly condition. Prometheus was a Titan, a descendent of the Sun, and one of the lesser Greek gods. Some say it was he who created the first man, taught him to walk erect, and gave him knowledge of all human arts. When Zeus, the ruling Greek god, decreed that human beings must eat their food raw, Prometheus took pity on the frail and vulnerable creatures and stole the divine fire from the heavens to improve the human state.

For this defiant act Zeus decreed that Prometheus must be fettered to a rock in desolate Caucasus forever.

In the time of Aristotle, the Greek philosopher, around 350 B. C., fire was considered one of the four essential elements of life and all matter. The other three elements were earth, water, and air.

Modern anthropologists believe that only in people's lowest state of barbarism were they without fire. If it was not Prometheus who gave fire to mankind, then all evidence indicates that it was lightning from the heavens that brought fire to earth. The vegetation of the open prairies and the trees of the forests must have, in various seasons and under varied conditions, fallen victim to fire induced by lightning. The bursts of flame at first must have terrified our earliest ancestors, but with the passing millenia they must also have recognized the benefits of warmth and light given off by fire. Through trial and error, perhaps with the aid of a log smouldering after the main fire had burned itself out, they put fire to their own use. Thus our ancestors took the first great step of many that raised the human species from savagery and barbarism to the achievement of civilization as we know it in the late twentieth century.

Archeological diggings show that mankind's first use of fire occurred around 500,000 B.C. Charred wood and bones have been found and dated to this era. Fire was used as a weapon and as a fuel for light, heat, and cooking. Thus it improved the life style of this period. Undoubtedly, it is one of the basic causes for the formation of the family unit, since it gave human beings the independence and self-sufficiency to move about and to relocate in groups in the colder zones outside the rain-forest belt.

By Neolithic times man and his mate had progressed to the point where they could create fire through some method of friction and need not carry live embers with them.

Around 7000 B.C., they were using fire in a slash-and-burn type of agriculture, burning off the brush so that the land could be tilled. (This new practice thus introduced the first ecological problem by the defoliation of the environment.)

The firing of pottery, the smelting of copper, the making of bronze from copper and tin were realities by 3000 B.C. The smelting of iron came two thousand years later — about 1000 B.C.

Step by step our ancestors made greater use of fire as a friend. The industrial revolution, with its smoking chimneys of factory and industrial plant, testified to the extent to which human beings had progressed with fire as their ally. Today's harnessing of atomic energy shows one of the latest uses of fire.

It is small wonder that fire has been venerated and worshiped. Its great energy, its similarity to the sun, its powers of light and destruction, its ever-changing volatility — all lend credence to the reverence in which it was held through many centuries.

Fire, as friend, has aided man in his long struggle to achieve the civilization of the late twentieth century. Fire, as foe, is the scourge from the sky that has leveled forest and city and that takes some 12,000 lives yearly in the United States alone.

Fire, friend and foe — technology still searches for further ways to harness this servant of the modern world. We still seek the knowledge that will help us to prevent loss of life and property because of fire.

The first record concerning either building safety or fire protection was written in the reign of Hammurabi in Babylonia. In the year 2000 B.C., he proclaimed that the builder whose structure collapsed because of faulty construction should die.

Probably the first step to provide a measure of fire protection was taken by the Roman Emperor Augustus. About 24 B.C., he instituted a night patrol of slaves, called a *vigile,* which checked for fires and alerted the town when fire was discovered. In England about A.D. 1000, the watch service became known as a curfew, from a French word meaning to cover or to extinguish fire by a certain nighttime hour. An ordinance established the hour at which all fires had to be extinguished. After that hour, a watch was maintained to enforce compliance and to check for any outbreak of fire.

The first written ordinance pertaining to fire safety was enacted by King John of England after the London fire of 1212. His decree provided for some protection against the spread of fire from building to building, the old problem of "exposure communication" that is still with us today.

In colonial America, the night patrol was termed a rattle watch, named for the clattering noise made by the rattler type of alarm device used by the watchmen (Figure 1-1). This noise alerted the sleeping town to a fire.

Years later, when an organized fire department had been established, the night watch worked from a tower on top of a fire station or from another high point in the community. When a fire was spotted, the watch noted its approximate location and aroused the fire company.

HISTORY OF FIRE ORGANIZATION

1

At the time of the Chicago Fire on October 8, 1871, Chicago had installed city fire-alarm boxes for reporting of fires by citizens, but a fire-lookout watch was still maintained. The initial response to the fire scene was initiated by a fire lookout. This tower watch service continued until more fire-alarm boxes and greater use of public telephones facilitated earlier fire detection and reporting. In a sense, the night watch continues today in the person of the police officer who patrols the streets all night and who many times turns in the first alarm.

Figure 1-1   Rattle used by night watchman to sound the fire alarm.

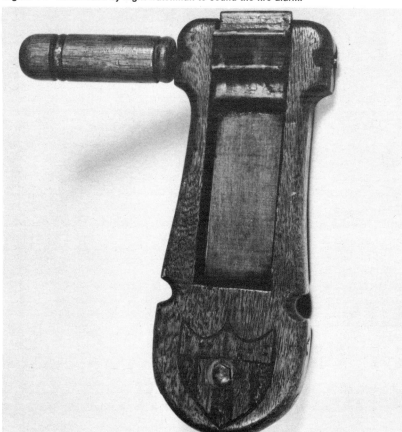

Historical Collection of the Insurance Company of North America.

DEVELOPMENT OF FIRE PROTECTION

The fire ordinance of King John was the first measure of fire *prevention.* The night watch service was the first step in fire *suppression.* Fire suppression techniques improved as technology expanded. Among the basic steps leading to modern firefighting practices were:

1. Use of leather buckets, swabs, and hooks

2. Development of the hand-tub water engine

3. Use of ladders in extinguishment and rescue

4. Development of hand-tub water engines with suction and hose lead-outs

5. Establishment of volunteer and paid fire departments

6. Development of hose carts

7. Introduction of the horse-drawn steamer pumper

8. Utilization of the gasoline engine and pumper

BASIC DEVELOPMENTS IN FIRE PREVENTION MEASURES

Early in our history, citizens were aware that they would be safe from fire only when fire is prevented before it starts. Basically, all fire prevention and fire-safety measures stem from three steps taken to prevent fire in this colonial period.

1. Initial building ordinances restricting the use of thatching as roof material and fireplaces with wooden, mud-plastered chimneys

2. Use of more comprehensive building codes incorporating increased fire- and life-safety features

3. Legislation granting legal power of enforcement and right of inspection

History details many examples indicating that reform or progress does not take place until a disaster makes corrective action mandatory. This is especially true in the fire protection field. Progress in fire suppression and prevention has been spurred by the conflagration—the fire that devastates large areas. (See Appendix A for more information on this type of fire.)

EARLY FIREFIGHTING TECHNIQUES—ENGLAND

A conflagration struck London in 1666. London was then the largest city in the world. On September 2, 1666, it was struck by a fire that raged out of control for five days. Some 13,000 homes were destroyed in addition to churches, warehouses, and other buildings.

Fire control in the early period, in both London and the American colonies, was handled by bucket brigades. Each household kept a container that could be used, together with others, in carrying water from the nearest source to the fire building (today's phrase for the building in which the fire occurs). Leather buckets were commonly used. At first, they were hand-sewn; later, they were riveted. In the colonies, when volunteer fire companies were formed, these buckets frequently bore a distinctive insignia or marking, a trademark of the volunteer company.

Two lines of volunteers were formed from the fire building to the source of water. One passed the filled bucket to the spot where it could be thrown on the burning building; the second line passed the empty buckets back to the water source to be filled. Hot embers beyond the range of the water throwers were smothered with water-soaked swabs on the ends of long poles.

There is some indication in history that the early Egyptians, Greeks, and Romans knew the principle of a pump that could be used to "throw" water. This principle seems to have been "lost" until the early sixteenth century when several men, working independently, designed and built models of "water engines." None of them had been put to practical use at the time of the London catastrophe.

This disaster spurred development and adoption of the model hand-tub water engines. Ladders also came into use for fire extinguishment and rescue work. Fire insurance was written for the first time to cover future fire loss.

The hand-tub fire engine in essence was a rectangular storage box topped by a smaller condensing case. Two pistons, operated by handles or "brakes," were connected to the condensing case. The brakes, placed at either the sides or the ends of the storage box, were large enough for volunteers to use them to supply the needed power for the pistons. Connected to the top of the condensing case was a water nozzle.

The bucket brigade supplied water to the storage box. The pistons then forced the water out through the pipe nozzle on top of the condensing case.

The water engine had to be placed quite near the fire building so that the stream from the fixed nozzle could reach the fire. However crude this method was in the light of later advances, it was a vast improvement over the more-or-less indiscriminate throwing of buckets of water in the general direction of the fire.

Prior to the London disaster of 1666, ordinances had been passed to stop the use of thatching as a roof material and the use of wooden, mud-coated chimneys. Neither ordinance appears to have been enforced. After the 1666 fire, a more complete code of building regulation was drawn up and passed, but an enforcement agency was not provided for until a hundred years later. Such a long delay seems inconceivable, yet even today funding and activation frequently lag far behind the legislative act.

EARLY FIREFIGHTING—THE COLONIES

The American colonists were confronted with fire soon after settling here. When one thinks of their type of dwelling, it is easily understandable why a serious threat of fire existed. The early settler's home was not the substantial log cabin typical of the American plains at a later period in our history. The exigencies of housing and lack of equipment and supplies forced the early colonists to construct low, small cabins with oval thatched roofs of straw and branches. The inside was dominated by a large stone fireplace with its own thatch or wood-constructed chimney. The chimney lining was plastered with mud. Most of the fires were started by hot embers falling on a dried-out thatched roof, or inside a chimney where dried mud had broken loose, exposing the wood underneath to the fire in the fireplace. The closeness of the houses, clustered together for security reasons, added the danger of communication of fire to the adjoining structures.

As mentioned previously, firefighting equipment consisted of the leather fire bucket, a swab for dousing burning embers on a thatched roof, and a hook attached to a length of rope for pulling down houses in the path of the spreading fire so that a firebreak could be made.

The causes of fire in those early times are analogous to the present ones: substandard housing, crowded conditions, lack of ordinances or lack of ordinance enforcement, and poor planning.

In January 1653, Boston experienced a fire of conflagration magnitude that resulted in three deaths and one-third of the families burned out. Once again, a city conflagration led to the passage of new fire protection laws:

1. Every house was required to have a ladder long enough to reach to the ridge pole—the ladder to be used for fire-fighting purposes.

2. Every house must provide a 12-foot pole with a swab to quench burning embers on the roof.

3. Six ladders and four iron crooks and chains were to be hung at the meetinghouse, readily available for any fire incident. The crooks and chains could be used to pull down houses to form a firebreak.

4. No house should be torn down for a firebreak without proper consent of the town authorities; when so ordered, the owner was to have no recourse.

5. A bellman was to be provided to patrol during the night hours.

6. A cistern for fire-fighting purposes was to be established at a prominent corner location.

7. No outdoor fire could be built within 3 rods of any building.

8. Anyone convicted of arson was subject to death.

9. Burning embers could not be carried from one point to another to be used as a "fire starter" except in a container that would prevent their being blown about.

10. No fires were permitted in ships docked at wharves.

News of the water engine and the new method of fighting fires in England soon reached America. Boston, in the year 1676, ordered one of these pumpers from England. Before its arrival, the city suffered another large fire, more serious than the fire of 1653. A third large fire in 1679, followed by a series of smaller fires, caused Boston to purchase two more engines in 1707. Another conflagration followed in 1711, destroying more than a hundred buildings. Three more water engines were purchased from England. Boston now had six engines, although New York and Philadelphia still had none.

It took the conflagration of 1730 to induce Philadelphia to order three of the new water engines. Two were the Newsham models made in London, and the third one was constructed locally by an Anthony Nichols.

In 1731, New York ordered two of the newer Newsham fire engines from London. Richard Newsham had improved the water engine and eliminated some earlier mechanical problems. His new model had two wooden pump handles at each side of the engine. These pumping handles were 15 feet in length, permitting many men to assist in pumping the water.

In 1743 an American-built fire engine went into service as New York's third such engine. The builder was Thomas Lote of New York. Quite possibly, the first American models borrowed many ideas from the Newsham model, but their designers, using American ingenuity and technology, soon introduced many refinements and characteristics of their own. For example, all engines were complete with brass trim, distinctive name, and individualized company markings. One of the larger American machines was John Agnews's "Southwark No. 38," capable of throwing a stream 180 feet. Some features of the improved machines are shown in Figure 1-2.

The next big improvement in the hand water-tub was the development of an engine that could draft its own water. Suction was taken from a well or cistern. This innovation, coupled with newly developed fire hose, meant that the engine could be at the main source of water and the nozzle could be removed from the top of the pump. With the nozzle now at the end of a movable length of hose, the volunteers could take a more advantageous position for fighting the fire. These improvements were made around the turn of the nineteenth century.

Figure 1-2 American-built hand water-tubs. (a) Sidestroke with foot treadle, manufactured by Richard Newsham, London, and imported to New York in 1731. (b) Endstroke model with education pipe and condenser case, built by John Agnew in1846.

Historical Collection of the Insurance Company of North America.

FIRE INSURANCE AND THE FIRE MARK

Fire insurance had first been written by London companies shortly after the Great London Fire of 1666. The colonies placed insurance with the same companies until the early 1750s. American-owned fire insurance companies then came into being.

Fire insurance in the colonies owes its beginning to Benjamin Franklin, philosopher, journalist, diplomat, and statesman. Franklin had become interested in firefighting during his youth in Boston, his birthplace. At seventeen years of age, he moved to Philadelphia. After founding his *Pennsylvania Gazette*, he began to write articles on fire causes and methods of extinguishment. He cofounded the first fire company, the Union Fire Company, in Philadelphia in 1736. In 1752 he became the cofounder of the first successful fire insurance company in America. The success of this company led to the establishment of many others.

The new fire insurance companies introduced another interesting sidelight to fire protection. It became the practice of the various insurance companies to place an insignia, a "fire mark," on the buildings they insured. Poured in lead or cast iron, these fire marks featured individualized and colorful designs. (Some of these marks are shown in Figure 1-3.) Many towns during this period rewarded the first company that put water on a fire building. Now the insurance companies formed or employed fire companies to protect their insured properties. The fire mark served a dual purpose. It gave visual evidence that the fire building was insured and identified the company that held the insurance—an inspiring sight for added efforts at the fire scene!

VOLUNTEER FIRE COMPANIES

The first volunteer company, Franklin's Union Company, was limited to thirty members. A second company, known as the Fellowship Fire Company, was formed in Philadelphia in January 1738. This company stored ladders at various key locations to be readily available when fire struck.

The concept of volunteer companies soon spread to other areas. (Figure 1-4 shows a New York firehouse built in 1790.) They became both social and business organizations. Competition between companies was keen. Drilling and spit-and-polish sessions were

Figure 1-3   Early fire marks. At first, fire marks were made of case or beaten lead. Some of them were taken to be made into bullets during the Revolution. Later marks were made of copper, zinc, iron, and, in rare cases, of terra cotta or stone which was then built into the wall of the insured property. The four insurance companies represented by these fire marks are still in existence today.

Figure 1-4   Replica of the 1790 firehouse of the Eagle Engine Company No. 13 in the Manhattan area of New York City.

Courtesy of the Firefighting Museum of the Home Insurance Company.

frequent. The desire to "steal" the fire from other responding companies led to rivalry that often was a deterrent to fire extinguishment. Many of these volunteer companies later formed fire insurance companies.

FURTHER DEVELOPMENTS IN FIRE PROTECTION

Rescue ordinances had required firefighting ladders and iron crooks or hooks to be placed at strategic locations. Hook-and-ladder companies were now formed. (See Figure 1-5.) At first, the hooks and ladders were carried to the fire on foot. Then came the development of a chassis long enough to carry this equipment. The fire truck also carried picks, axes, leather buckets, whale-oil torches, and firehats. It responded to the fire scene along with the pumping engine.

Figure 1-5 Early hook-and-ladder trucks: The first hook-and-ladder companies had no apparatus, but carried their hooks and ladders to the fire on foot. The development of a chassis long enough to carry this equipment marked another step forward in fire protection. An early hook-and-ladder truck carried ladders, picks, axes, leather buckets, whale-oil torches, and hats.

Historical Collection of the Insurance Company of North America.

Figure 1-6 A typical wooden main of 2 3/4-inch diameter used in the installation of early underground systems. This particular pipe, installed around 1842, was recovered in Chicago in 1967 by electrical workers installing an underground service.

In the early 1800s, water mains were installed to serve domestic and fire needs. (See Figure 1-6.) The first mains were pine logs with a 3- or 4-inch hole bored through the center. There were no hydrants, but a removable wooden plug (hence the term "fire plug") was installed at various points. These plugs could be removed to obtain water for fire control. As pressures in the mains increased, fire hydrants were installed. They led to the establishment of hose-carrying apparatus to accompany the fire engine company. This hose was used for lead-outs to the fire or for supplying water to another engine in line.

THE STEAM FIRE ENGINE

The first steam-driven fire engine was imported from England in 1829, but the volunteers felt it was a direct affront to their ability. Therefore it was not used. A model made by a Paul Hodge for New York was also put in service but not used.

It was the city of Cincinnati, with a steam engine built by the Latta brothers of that community, that forced the use of the steam engine in the United States. This apparatus could get up steam in four minutes and throw a stream 300 feet. It required four horses to pull it to the fire scene. The hand water-tub had ended the era of the bucket brigade. Now the colorful and breathtaking sight of fast-running volunteers drawing the hand water-tub down the street, often in competition with another volunteer company, was replaced by the even more pulse-quickening sight of racing horses pulling a smoke-belching steamer down the roadway.

The era of steam was in full sway. The galloping horses of the stage-coach lines were replaced by the steam railroad. The sailing ship was replaced by the steamship. Steam now generated energy previously supplied by both human and animal physical power.

The control of fire had led to harnessing the power of steam; now steam was to be used to extinguish the uncontrolled fire. The steam fire pumper (Figure 1-7) replaced the eighty men required to operate the hand water-tub. (In action at the fire scene, the crew operating the brake of the hand tub had to be replaced at frequent intervals.) Now the chugging steam pumper was capable of supplying four lines.

Figure 1-7 Early steam pumpers.

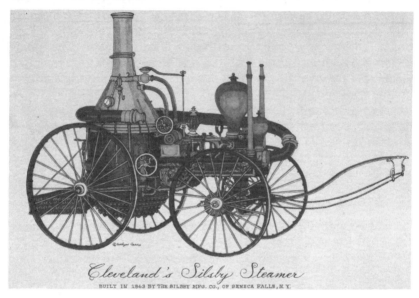

Cleveland's Silsby Steamer

BUILT IN 1863 BY THE SILSBY MFG. CO., OF SENECA FALLS, N.Y.

Philadelphia's Hurricane 13

A UNIQUE AND BEAUTIFUL ENGINE, BUILT IN 1860 BY G. J. AND J. L. CHAPMAN OF PHILADELPHIA.

Courtesy of Ridgemoor Archives, Chicago.

But man and technology did not stand still. If the nineteenth century was the age of steam, then the twentieth century is the age of electricity and the gasoline engine.

Despite the never-to-be-forgotten thrill of seeing a horse-drawn steam fire pumper racing to a fire, this method had its handicaps. When responding to an alarm, a fire to raise the necessary steam immediately had to be started to ensure operation on arrival at the fire scene. Sparks from this fire offered a fire hazard of its own while in route. Response was slow: Horses had to be hitched up and they could not travel as swiftly as the gasoline driven vehicle.

Early in the 1900s, it became apparent that the steam-operated horse-drawn fire apparatus would go the way of the bucket brigade and the hand tub. The gasoline engine had been produced and was on the streets. It was only a question of time until this engine would propel the fire apparatus and operate the fire pumps at the fire scene. By the early 1920s most steam apparatus had been replaced by the gasoline engine pumper.

Building-construction techniques also were progressing. Buildings were reaching toward the "skyscraper" heights of ten stories. A manually raised aerial ladder had been placed in service in San Francisco in 1870, designed by a member of the San Francisco Fire Department to reach increased heights. Spring-assisted 85-foot aerial ladders were built in 1905. The mid-1930s brought the development of the power-raised 100-foot aerial ladder, now either of wood or metal construction.

To aid in getting water to the main seat of fire in upper-story fires, water towers were developed. In turn, they have been largely replaced by elevated platforms. The elevated platform, or "snorkel" unit, also can be used as a rescue unit in particular instances.

Although water is still the best and cheapest extinguishing agent, its use has been upgraded and made more sophisticated. Pumper capacities have increased to 1,500 and 2,000 gallons per minute (gpm). Most pumpers now carry booster tanks with a 300- to 750-gallon water supply that can be used, usually with 1-inch hard-rubber booster lines, for "instant" water at the fire scene. Fire suppressants or retardants added to the water can aid in extinguishment. "Slippery" water, an additive that reduces friction loss, increases gallons-per-minute flow or permits the use of hose of smaller diameter.

CHEMICALS AS EXTINGUISHING AGENTS

Chemicals used as extinguishing agents were introduced into the fire service in 1873. The need for getting water on the small, incipient fire as soon as possible to prevent a larger fire was apparent. The early solution to this problem was a chemical engine, a small horse-drawn apparatus carrying 100 gallons of water mixed with bicarbonate of soda.

Upon arriving at a small fire, a firefighter dumped sulfuric acid into the water solution. In brief, in the chemical reaction of the bicarbonate of soda and sulfuric acid, carbon dioxide gas was formed which expelled the water through a 1-inch line to the seat of the fire. These chemical engines became standard equipment for use on small fires and in outlying areas that had water-shortage problems.

Now, many chemicals are used in either fixed installations or mobile units, usually for a specific hazard. Dry chemical units, CO_2 units, various types of foams including "light" water, and the halogen agents are used in apparatus varying from a small portable extinguisher to a large airport crashwagon.

The local fire department ranges in size from a manpower pool of a few volunteers serving a rural area to a large metropolitan force of over 1,000 fully paid employees. But most of the suburban departments, having only a few full-time paid firefighters, still depend entirely or heavily upon the volunteers.

SUMMARY

The history of the fire service in the United States parallels the history of the country. Faced with still another hazard in their environment that challenged their existence, citizens rose to the challenge. They adapted existing technology to aid in their battle against fire. From leather buckets, swabs, and hooks, they progressed to a "fire engine" that could propel the water to the fire by means of a fire nozzle. The moving force was manpower.

As new technology was developed, it was modified to serve the needs of the fire service. In some cases, technological advances peculiar to the needs of the fire service were effected. Ladders were used in extinguishment and rescue. Volunteer fire companies were formed. The hand water-tub engine was improved so that it could take suction. Hose was developed to bring the nozzle and the water to the most advantageous point of attack. The use of steam power was applied to the steam fire engine. Later, the gasoline engine became the power for the fire pumper. Ladder trucks and specialized equipment were developed.

Laws and ordinances were formulated and enacted not only to curb existing hazards but to prevent their occurrence in the future.

DISCUSSION TOPICS—DEVELOPMENT OF FIRE PROTECTION MEASURES

1. Cite and discuss examples of early fire-watch services.

2. List some important steps in the development of fire-fighting techniques.

3. Discuss the importance of legal powers of enforcement and right of inspection in fire prevention.

4. Name two fires of catastrophic proportion and discuss resulting changes in fire prevention and suppression measures.

5. What important developments in firefighting took place in the late 1790s and early 1800s?

6. Why was the steam fire engine a big improvement in firefighting?

RESEARCH PROJECTS

1. Research the history of an early fire company—its organization, training, fire-fighting techniques, and other characteristics.

2. Research the formation of an early fire insurance company. Discuss the role insurance companies have played in fire protection.

3. Outline the parallelism between technological advances in societal living and fire service practices. Include your concept of present or potential advances in our life style and their adaptation to the fire service.

4. It has been said that it takes a catastrophe to bring needed change. Outline some changes in fire protection measures resulting from some serious incident of fire.

FURTHER READINGS

Bulan, Alwin E., *Footprints of Assurance,* Macmillan, New York, 1953.

Dunshee, Kenneth Holcomb, *Engine!–Engine!,* Home Insurance Co., New York, 1939.

Hayward, Charles F., *General Alarm,* Dodd, Mead, New York, 1967.

McCosker, M. J., *The Historical Collection of Insurance Company of North America,* Beck Engraving Co., Philadelphia, 1967.

Morriss, John V., *Fires & Firefighters,* Random House, New York, 1955.

The fire service and those concerned with fire safety cannot control fire until they are aware of where it is striking and its basic causes. These facts cannot be known unless accurate and meaningful statistics are kept of fire responses. If the causes and effects of fire are known, remedial measures can be taken through fire prevention and building code ordinances to curtail it. The data must be meaningful. It takes both a combustible material and a source of ignition to cause a fire, and both should be identified in the fire report. But even more information should be included if the report is to prove useful in curtailing other fires of a similar nature.

FIRE-INCIDENT REPORTING

The National Fire Protection Association, at the present time, collects and publishes national statistics on fires and fire losses. It is the NFPA that stands in the forefront of those who advocate that more meaningful data be gathered. To that end, the NFPA has published guidelines on the data that should be incorporated into a fire-incidence report. Under these guidelines, the following information would be included:

1. *Area of origin:* This pinpoints the fire to a particular room, portion of a room, or object within the structure or open area.

2. *Equipment involved in ignition:* The source of heat for ignition may be attributed to equipment that gives off or uses heat. For example, cooking equipment or electrical appliances.

3. *Form of heat of ignition:* The actual heat of ignition will be in a particular form or manner of a particular fuel or heat energy source, such as smoking materials, an open flame, or a spark.

FIRE STATISTICS

4. *Type of material ignited:* This item identifies the particular class of combustible material ignited—say, flammable or combustible liquid, wood, or paper.

5. *The form of material ignited:* This item includes the form and use of the material ignited (such as roof covering or furniture).

6. *Act or omission:* This includes human acts, or omission of action, either accidental or deliberate.

These main listings may be further subdivided so that all sources or possible combinations of causes can be effectively tabulated.

A good report will also include several other factors pertaining to the building proper that may have contributed to the original fire or extension of fire:

1. *Occupancy or nonoccupancy:* What particular use (residential or other) has the building been put to, and did this use play a role in this particular fire?

2. *Type of construction and materials used (such as frame or brick):* Both these factors may have a bearing on the fire.

3. *Floor elevation of initial origin:* The elevation of the fire floor often indicates the difficulty of, and special techniques possibly used in, extinguishment. (Fires below grade, such as a basement fire, and high-rise fires require different treatments.)

These facts, in sufficient detail, should supply enough information to properly classify building fire causes and building fire losses.

A record should be kept of each fire response. In addition to information on the causes of each fire, the fire report should record:

1. Location of fire

2. Owner's name and address

3. Fire units responding

4. Extent of fire

5. Method of extinguishment

6. Damage estimate

7. Weather conditions

8. Life hazard (injuries or death)

(The elements of a good fire report, together with an example, are discussed in Appendix B.)

There is a growing tendency in the fire service to use electronic data processing equipment to tabulate fire reports. The types of information listed here can be easily coded for storage and retrieval for future analysis. A periodic retrieval, reviewing, and analysis of the stored data will indicate trends and areas of concern to the municipality or fire district.

The National Fire Protection Association has incorporated a list of all the information it deems necessary to a reliable fire report into NFPA Bulletin 901, *Coding System for Fire Reporting* (1971).

Proper data properly analyzed can supply the fire department with the following pertinent information:

1. Any increase in the number of fire responses and its possible relationship to changes in population density and home repair and maintenance breakdown.

2. Construction types and occupancies involved in fire and their correlation with the fire prevention program.

3. Fire response by time of day, season, and specific weather conditions, and prediction possibilities of future fires.

4. Fire response and its relationship to equipment location and manpower.

Each municipality or fire district is responsible for gathering its own statistics. Such information is essential if the necessary fire defenses are to be established and the public is to be alerted to the problems. When the statistics are incorporated into a national fire-loss survey, they can be of value to all fire departments and other organizations concerned with fire safety.

NATIONAL STATISTICS ON FIRE LOSSES

At the present time, national statistics on fires and fire losses, published by the National Fire Protection Association, are based on statistics and estimates gathered by the NFPA's record department. The association emphasizes that the published statistics are estimates only, based on its limited statistical survey and educated "guesstimates" from its years of experience in the field.

In the compilation of the data, the NFPA relies upon the following sources:

1. Reports of fifteen selected state fire marshals

2. Detailed reports of sixteen city fire departments in states that do not issue state fire-marshal reports

3. General data submitted by 726 large cities on fires by occupancy

4. NFPA files on large loss fires

Other sources, such as the U.S. Forest Service and the U.S. Coast Guard, provide information on specific classifications of fire.

The published data indicate changes in fire causes and losses. In the course of years, coal-furnace and hot-ashes fires have declined, owing to greater use of gas and oil as heating fuels. The replacement of the live green Christmas tree by the artificial tree has substantially reduced the Christmas tree fire. The artificial tree, however, has presented new problems; when subjected to fire, the synthetic fibers may contribute to the combustion process and, in addition, may give off noxious fumes. The metal limbs are a potential short-circuiting agent where electric lighting is used.

The data also reveal trends of fires in new products, new services, and new processes. These trends emphasize the value of technological advances in firefighting and the importance of a greater understanding, through education, of the causes of fire. Only with this knowledge can the new fire hazards be handled properly by the fire service.

The annual loss of life through fire in the United States hovers at the 12,000-victim mark. Of these victims, approximately 6,500 lose their lives while in the supposed safety of their own homes, and one-third of them are children! In addition, fire causes several hundred thousand nonfatal injuries yearly.

The victims suffering either fatal or nonfatal injury may be classified into three broad categories:

1. The victim was close to the burning material and suffered burns—possibly fatal. In most cases, it is clothing, bedding, or upholstered furniture that is on fire, but flammable liquids or gases that cause an explosion or flash fire may engulf the victim.

2. The victim cannot escape from the fire scene. Exitways are blocked by fire, smoke, heat, and gases which quickly fill the fire building. The victim usually suffers no burns. Rather, he or she is injured, perhaps fatally, by what is generally called smoke inhalation.

3. The victim may not have been a building occupant, but may have entered the fire area for fire-fighting or rescue purposes.

It is the multiple-death fire that usually makes the front page of the local or syndicated newspaper. The airplane crash, the coal-mine disaster, a nursing home holocaust make headlines. These tragedies are outweighed by the cumulative total of fire victims in the unspectacular fire, usually in the residential type of structure. Even in the multiple-death category (three or more victims), the residential fire is responsible for the greatest loss of life. (See Table 2-1 for a breakdown on multiple-death fires.) In 1973, of 205 fires in the United States in which there were three or more victims, 701 of the 1,008 fatalities lost their lives in 184 residential fires.

The annual dollar loss to the nation in physical assets alone reached an estimated $2.927 billion in 1972. This annual loss occurred in an estimated 2½ million fires.

MAJOR CAUSES OF FIRE

An examination of fire reports quickly indicates the large number of fires attributable to human beings in their handling of those combustible materials and ignition sources that have caused the fires. Improved equipment has led to quicker extinguishment. Stricter codes and better materials have made buildings more fire-resistive. Detection and fixed extinguisher systems lead to early extinguishment. Yet almost a million fires occur yearly!

To obtain statistics on the causes of fire and the dollar loss in each category, classifications of fire causes have been established. Building losses by cause are shown in Table 2-2. The most frequently listed causes of fire merit discussion here.

Table 2-1 Multiple-Death Survey Chart, 1973

	Number of Multiple-Death Fires	Number of Deaths
Total	205	1,008
Building fires		
Residential		
One and two-family dwellings	94	359
Apartment houses	48	235
Mobile homes and trailers	13	52
Other residential buildings	10	55
Total	165	701
Institutional		
Rest and nursing homes	5	33
Other institutional buildings	1	3
Total	6	36
Schools	1	3
Industrial, commercial, and storage buildings	4	25
Public assembly	4	45
Other	4	52
Total	184	862
Nonbuilding fires		146
Transportation		
Aircraft*	6	69
Water transport	5	25
Road transport	6	25
Other transport	3	22
Total	20	141
Other nonbuilding fires	1	5
Total	21	146

*Subject to change when final reports are made available by the National Transportation Safety Board.

Source: Reprinted from *Fire Journal*, May 1974, copyrighted by the National Fire Protection Association. Reprinted by permission.

Table 2-2 Estimated Building Fire Losses by Cause, United States, 1972

These estimated figures are intended to show the relative order of magnitude of fire losses by cause, and to indicate year-to-year trends. While they are reasonable approximations based on experience in typical states, they should not be taken as exact records for each class. The figures by themselves do not show the relative safety in use of various types of materials, devices, fuels, or services, and they should not be used for that purpose. Reproduction of this table, in whole or in part, is authorized only with written permission from the Association and with the following identification of figures: National Fire Protection Association estimates.

CAUSE	NUMBER OF FIRES		ESTIMATED LOSS	
Heating and cooking equipment				
Equipment, defective or misused	89,400		$116,700,000	
Chimneys and flues overheated or defective	21,800		16,200,000	
Hot ashes and coals	6,800		3,900,000	
Combustibles near heaters and stoves	37,200		40,800,000	
Total		155,200		$177,600,000
Smoking-related		109,700		95,900,000
Electrical				
Wiring and general equipment	101,600		203,100,000	
Motors and appliances	61,000		112,700,000	
Total		162,600		315,800,000
Rubbish fires		36,000		2,400,000
Flammable liquids (excluding those associated with heating and cooking equipment)		65,200		56,900,000
Open flames and sparks				
Sparks and embers	6,200		6,700,000	
Welding and cutting	8,200		28,800,000	
Friction, sparks from machinery	17,000		22,000,000	
Thawing pipes	5,500		11,900,000	
Miscellaneous open flames and sparks	35,000		32,800,000	
Total		71,900		102,200,000

Table 2-2 *(continued)*

CAUSE	NUMBER OF FIRES	ESTIMATED LOSS
Lightning	22,700	$ 43,300,000
Children and matches	69,200	74,600,000
Exposure	25,400	23,400,000
Incendiary and suspicious	84,200	285,600,000
Spontaneous ignition	15,100	25,900,000
Gas fires and explosions not included above	8,700	23,400,000
Explosions from fireworks, explosives, etc.	4,200	5,200,000
Miscellaneous known causes	65,900	191,400,000
Unknown	154,200	992,700,000
Totals	**1,050,200**	**$2,416,300,000**

Source: Fire Journal, September 1973, copyrighted by the National Fire Protection Association. Reprinted by permission.

Smoking and Matches

Careless smoking and disposal of smoking materials, most often cigarettes and matches, usually rank as the largest single fire cause. Smoking is also responsible for the greatest loss of life by fire. The careless disposal of cigarettes and matches accounts for the highest number of fire-burned victims in the home.

The potential for fire is great because of the continued burning of the discarded cigarette. The major contributor is the smoker who falls asleep while smoking in bed or in an upholstered chair. It usually requires some time before the smouldering cigarette causes ignition temperatures to be reached in the mattress or upholstery. By this time the smoker is soundly asleep and can be overcome by the gases and smoke generated.

Cigarettes and matches are also responsible for many fires in the forest, along the roadside, and in other areas where smoking is permitted.

Electrical Fires

Ranking No. 1 in 1972 as a cause of fire are electrical hazards. This category can be broken down into the hazards of fixed installations and those of power appliances.

The electrical wiring, panel boards, fuse boxes, and current breakers constitute fixed installations; they are subject to short circuits, arcing, and sparking that are a potential source of fire. Those appliances that use electricity as a source of power, such as lamps, irons, and power tools, may, through improper use, poor maintenance, or faulty installation, contribute to our national statistics on fire.

Heating and Cooking Fires

A third major cause of fire are heating and cooking equipment and processes. This category can be further subdivided.

DEFECTIVE AND OVERHEATED EQUIPMENT

Equipment is the largest source of heating- or cooking-equipment fires. Again, the human element, rather than basically poor equipment, is to blame. Faulty installation or the poor maintenance of furnace, heating unit, or domestic or commercial oven and range is primarily responsible.

CHIMNEY AND FLUES

Flue linings that are cracked and connections that have separated are possible ignition sources. Soot that has been permitted to build up in the flue can be heated to its ignition temperature. Combustibles placed too close to the flue may be set afire.

HOT ASHES OR COALS

Hot ashes in combustible containers, or stored near combustibles, have long contributed to the heating and cooking fire. Over the past years, with the lessened use of coal as a source of heat, this cause has decreased, but it is still a frequent culprit in the United States.

COMBUSTIBLES NEAR HEATERS

Storage of combustibles, such as clothing, furniture, or rubbish and trash, too near to the heating source is responsible for a considerable number of fires. These fires, although placed in the heating and cooking classification, are not the fault of defective equipment or poor installation or abuse of the equipment, but again, are caused by carelessness.

Children and Matches

Matches hold a special, and sometimes fatal, fascination for children, as attested by the number of children who set their clothing on fire. Terrified, they then run—thus fanning the flames. Many cases can also be cited of the child who accidentally starts a fire in one room while playing with matches, then flees into another room, afraid to reveal the fire he or she has started for fear of being scolded and punished.

As the records show, men and women and human error still contribute heavily to the loss figures. Is complacency to blame? Or lack of knowledge as to what constitutes a hazard?

Richard E. Bland, chairman of the National Commission on Fire Prevention and Control, summarized the problem this way in *Fire Journal* magazine, July 1972:

> If one prevents fire he has neither to detect it nor to extinguish it. But here is a people problem, one for the soft sciences. Is the population apathetic? Or is fire of such personal consequences that concern is unconsciously suppressed? If the population is apathetic to fire threat, how do we best proceed toward awareness? If the obstacle is horror suppression, how then do we condition our associates to face that threat with confidence that destructive fire can be effectively controlled?

FIRE LOSSES BY OCCUPANCY

Classification of the national fire loss by occupancy is also important. Here, too, for the purpose of easier statistics gathering, broad categories of occupancies have been listed. Within each classification are inherent hazards peculiar to that classification. Not only do the statistics indicate where most of our fires are occurring, but they suggest where efforts need to be expended in developing fire protection and fire prevention methods to reduce these losses.

The occupancies (excluding storage and nonbuilding occupancies) with the largest number of fires, in rank order, are:

1. Dwellings, one- and two-family

2. Apartments

3. Trailers and mobile homes

4. Restaurants and taverns

Fire losses by type of occupancy are listed in Table 2-3. The table shows that the largest dollar losses occur in the manufacturing and mercantile establishment fires. These are the "spectacular" fires; the fires that require much equipment and that receive detailed notices in the local papers. The largest losses usually are in an occupancy or process that is highly technical or complicated. It is here that the services of a skilled fire-safety engineer are needed to evaluate the hazards and to come up with the recommendations that, it is hoped, will eliminate future fires. The average firefighter generally does not have the training or qualifications to perform these tasks.

The more common, less complicated dwelling fire is another story. Table 2-3 indicates quite clearly that the dwelling fire is far in the lead in the number of fires and total monetary loss.

It is well within the realm of the firefighter to evaluate the causes of fire in the average dwelling, or in the three top-ranking occupancies shown in Table 2-3. He also can make recommendations to reduce the small, unspectacular fires that still are responsible for over 50 percent of the annual loss of life by fire, and that, as a category, account for the greatest monetary loss. The firefighter's capabilities in this regard point up the need for, and value in, fire prevention work and preplanning inspection on the company level.

The majority of fire calls are for nonstructural fires: a fence, rubbish, prairie, an automobile, or a deliberate false alarm.[1] These legitimate nonstructural fires can usually be handled with a hand pump or booster line.

Most building, or structural, fires will be extinguished with minimal use of water, 30 to 100 gallons per minute (gpm) usually being ample. Property damage and loss also are reasonably small. However, property loss must be quite high before the fire is actually ranked as a large-loss fire.

Large-loss fires as classified by the NFPA are those fires that cause a loss of $250,000 or more. The large-loss fire statistic again shows the large dollar loss that occurs in the manufacturing and the mercantile occupancy classification. Roughly, 25 percent of large-loss fires are in manufacturing occupancies and 25 percent in mercantile occupancies. The proportion of these dollar losses in the actual total dollar loss also follows the 25 percent figure rather closely. (See Table 2-4.)

[1] Recent statistics indicate that one of three, or at most one of four, fire calls is a deliberate false alarm.

Table 2-3   Estimated Fire Losses by Occupancy, 1972*

These estimated figures are intended to show the relative order of magnitude of fire losses by occupancies. While they are reasonable approximations based on experience in typical states, they should not be taken as exact records for each class. Any reproduction of these figures should be identified as follows: National Fire Protection Association estimates.

OCCUPANCY	NUMBER OF FIRES	ESTIMATED LOSS
Public assembly occupancies		
Amusement centers, ballrooms	2,500	$10,600,000
Auditoriums, exhibition halls	700	5,600,000
Bowling establishments	900	9,300,000
Bus, train terminals	600	2,600,000
Churches	4,300	28,100,000
Clubs, private	3,400	14,200,000
Restaurants, taverns	21,700	54,300,000
Theaters, studios	1,200	13,400,000
Other public assembly occupancies	2,600	15,100,000
Total	37,900	$153,200,000
Educational occupancies		
Schools, through twelfth grade	17,200	76,100,000
Colleges and other schools	5,200	14,800,000
Total	22,400	90,900,000
Institutional occupancies		
Homes for the aged, etc.	6,100	3,900,000
Hospitals	10,500	12,200,000
Mental institutions	800	1,500,000
Other institutional occupancies	3,800	7,200,000
Total	21,200	24,800,000

OCCUPANCY	NUMBER OF FIRES	ESTIMATED LOSS
Residential occupancies		
Apartments	109,000	$151,600,000
Dwellings, one- and two-family	562,500	638,500,000
Hotels	11,200	25,100,000
Motels	5,200	18,500,000
Rooming and boarding houses	5,000	17,200,000
Summer cottages and camps	5,500	11,400,000
Trailers and mobile homes	27,400	42,000,000
Other residential occupancies	9,800	14,100,000
Total	735,600	$918,400,000
Mercantile and office occupancies		
Appliance and furniture stores	4,100	28,100,000
Clothing stores	4,500	21,800,000
Department and variety stores	4,600	41,900,000
Drugstores	2,900	11,800,000
Grocery stores and supermarkets	6,900	36,900,000
Motor vehicle sales and repair facilities	9,700	35,100,000
Offices and banks	16,100	48,700,000
Service stations	5,400	11,600,000
Other mercantile occupancies	22,700	141,800,000
Total	76,900	377,700,000
Basic industry, defense, and utility occupancies		
Electric power plants	$3,100	$24,700,000
Laboratories and data-processing centers	800	2,800,000
Mines and mineral product plants	1,600	44,500,000
Nucleonics facilities	100	1,300,000
Other basic industry occupancies	1,400	8,300,000
Total	7,000	81,600,000

Table 2-3 *(continued)*

OCCUPANCY	NUMBER OF FIRES	ESTIMATED LOSS
Manufacturing occupancies		
Beverage, tobacco, and essential oil plants	900	$ 6,700,000
Drug, chemical, paint, and petroleum plants	3,800	94,900,000
Food product plants	3,700	42,200,000
Laundry and dry cleaning plants	3,400	9,900,000
Metal and metal product plants	4,100	54,000,000
Paper and paper product plants	3,000	11,800,000
Plastic and plastic product plants	1,900	16,500,000
Printing plants	1,600	6,400,000
Textile and textile product plants	3,500	18,100,000
Wood and wood product plants	3,100	46,500,000
Other manufacturing occupancies	12,000	82,000,000
Total	41,000	$389,000,000
Storage occupancies		
Barns, stables	19,300	81,000,000
Bulk plants, tank farms	1,500	10,300,000
Garages, residential parking	26,000	30,500,000
Grain elevators	2,400	42,800,000
Lumber, building materials storage	1,300	20,700,000
Sheds, farm storage buildings	14,000	30,100,000
Other storage buildings	10,400	111,800,000
Total	74,900	327,200,000
Miscellaneous buildings not listed above	33,300	53,500,000
Total building fires	**1,050,200**	**$2,416,300,000**

OCCUPANCY	NUMBER OF FIRES	ESTIMATED LOSS
Nonbuilding occupancies		
Aircraft, aerospace vehicles	200	$ 198,000,000
Crops in field	22,000	29,000,000
Forests	125,000	128,000,000
Grass, brush, rubbish	989,900	
Motor vehicles	550,300	127,300,000
Ships, railcars, etc.	20,000	29,200,000
Total nonbuilding fires	**1,707,400**	**$ 511,500,000**
Total fires	**2,757,600**	**$2,927,800,000**

*In addition to the fire-related incidents listed, fire departments responded to 860,000 deliberate false alarms.

Source: Fire Journal, September 1973, copyrighted by the National Fire Protection Association. Reprinted by permission.

FACTORS CONTRIBUTING TO THE LARGE-LOSS FIRE

These large-loss fires, with only a few exceptions, began as small fires that could have been easily extinguished. Other factors not present in the small-loss fire played an important role that permitted the incipient fire to increase greatly in magnitude.

Some factors that enter into the small fire's becoming a large-loss fire deserve attention here.

Building Location

In recent years industry has shown a growing tendency to locate far from the metropolitan area. The reasons include land costs and taxes. In some instances, such locations have proved to be false economy. The plant is far from the response area of a well-staffed fire department. As will be shown later, response time is a critical factor in any fire call. Response time can also be lengthened when the building is located in a comparatively inaccessible area of the city; in an area that is subjected to train or river bridge blockage; in a section that has poor or heavily traveled roads that slow response.

Table 2-4   Large-Loss Fires, 1973
A   Occupancies Where Large-Loss Fires Occurred, 1973

OCCUPANCY	NUMBER OF LARGE-LOSS FIRES	LOSS	NUMBER OF LARGE-LOSS FIRES	LOSS
Public assembly				
Bowling establishments	5	$ 2,310,000		
Churches	12	6,060,000		
Clubs	11	5,041,000		
Restaurants, night clubs and taverns	26	15,688,000		
Other public assembly places	5	2,776,000		
Total			59	$ 31,875,000
Educational				
Non-residential schools	24	17,878,000		
Other educational	3	2,000,000		
Total			27	19,878,000
Institutional			2	1,100,000
Residential				
Apartments	25	12,531,000		
Hotels and motels	14	11,388,000		
Other residential	8	3,816,000		
Total			47	27,735,000
Mercantile				
Food sales	13	7,363,000		
Textile product sales	10	5,603,000		
Household goods sales	10	5,021,000		
General item sales	14	6,362,000		
Offices	22	25,016,000		
Other commercial	38	19,035,000		
Total			107	69,400,000
Basic Industry				
Utilities	5	5,512,000		
Other basic industry	9	5,173,000		
Total			14	10,685,000

OCCUPANCY	NUMBER OF LARGE-LOSS FIRES	LOSS	NUMBER OF LARGE-LOSS FIRES	LOSS
Manufacturing				
Food processing	22	$23,097,000		
Wood and wood paper products	34	19,515,000		
Chemical, plastic and petroleum products	26	25,813,000		
Metal and metal products	23	26,457,000		
Other industrial and manufacturing	35	41,999,000		
Total			140	$136,881,000
Storage				
Agricultural products	20	13,400,000		
Textile products	8	8,730,000		
Wood and wood paper products	28	19,988,000		
Chemical, plastic and petroleum products	15	14,734,000		
Metal and metal products	16	18,556,000		
General items	14	10,790,000		
Other storage	11	37,878,000		
Total			112	124,076,000
Other Occupancies				
Outdoor properties	3	3,700,000		
Unoccupied properties*	38	29,078,000		
Ships and other water vessels	4	3,500,000		
Rail vehicles	4	10,500,000		
Road vehicles	5	3,306,000		
Aircraft	6	14,500,000		
Total			60	64,584,000
Total			**568**	**$486,635,000**

*Includes buildings under construction, renovation, and demolition.

Table 2-4 *(continued)*
**B   1973 Large-Loss Fires by Occupancy Class
(United States and Canada)**

OCCUPANCY CLASS	PERCENTAGE OF LARGE-LOSS FIRES	PERCENTAGE OF LOSS IN LARGE-LOSS FIRES
Public assembly	10.4	6.5
Educational	4.8	4.1
Institutional	0.4	0.2
Residential	8.3	5.7
Commercial	18.7	14.3
Basic industry	2.5	2.2
Industrial and manufacturing	24.5	28.2
Storage	19.8	25.5
Special occupancies	10.6	13.3
	100.0%	100.0%

Source: Fire Journal, July 1974, copyrighted by the National Fire Protection Association. Reprinted by permission.

Building Construction

Building construction will be discussed in greater detail in Chapter 7. Yet, without going into specifics at this time, it is obvious that building construction can contribute to the spread of fire. The wood-frame or wood-joisted building and the large building with few horizontal separations and many vertical openings are much more prone to the large-loss fire than the fire-resistant structure. (It is an interesting sidelight that the largest dollar loss of any sizable building fire in the United States—$52 million—occurred in a fire-resistant building in Chicago in 1967—the McCormick Place Exhibition and Convention Hall. The responsibility for this fire, however, falls more directly on careless housekeeping.)

Poor Housekeeping and Maintenance

An accumulation of rubbish and débris, excessive stock piling, and open fire doors are all invitations to a large-loss fire. The combustibles present are usually of a low ignition temperature: paper

ignites at 425° F, wood at 500° F or higher, and the added fire potential of rubbish or excessive stock contributes to the resultant high
loss. It was the débris and rubbish resulting from the removal of an
exhibition show at Chicago's Convention Hall that intensified the
heat and hastened the spread of fire that totally destroyed the
fire-resistant structure.

Poor building maintenance, such as broken windows, dry rot,
unpainted wood, and faulty electrical and heating installations, can
also cause or quicken the rapid spread of fire once ignition has been
initiated.

Delayed Discovery

Many plants employ around-the-clock watchmen on the premises. Others have internal alarm systems connected to watch services or to the local fire department. In these cases, the advent of fire
usually is quickly detected and reported. Most large-loss fires are
experienced either over the weekend when the plant or business is
shut down, or during the normal workweek between the hours of 8
p.m. and 8 a.m., where no watchman or watch service exists. Detection is usually made by the transient passerby, and then only when
the fire has reached sufficient proportion to be visible from the
exterior and perhaps at a distance.

Delayed Alarms

In those cases of large-loss fires which occur during business
hours, it is often determined that discovery of the fire was made
during its initial phase but that, before reporting the fire to the local
fire department, the building occupants took time to fight the fire
with the available extinguishing means at hand. Only when they saw
that the fire could not be extinguished by these means was the fire
department called. In other cases, a superior had to verify the fire
before the alarm could be given to the fire department. As will be
shown in our later study of time and temperature factors in the
spread of fire, prompt notification to the fire department is crucial. If
extinguishment means at hand can then extinguish or control the
spread of fire until the arrival of the fire companies, property will be
saved—but with the back-up of the fire department if needed.

Water Supply

Water supply can be affected by the building location. Some areas have a poor water supply. Hydrants may be few and far between or on small-sized mains incapable of supplying the amount of water needed. In some areas hydrants are nonexistent and the water supply must be taken from rivers or ponds or brought in by a water-tank truck. In very cold weather, the possibility of frozen hydrants, ponds, or streams can greatly handicap the firefighter's work. Again, the delay partially accounts for the time loss and temperature rise in the resultant large-loss fire. Hot weather can also have its adverse effects where water supply is limited. At best, the greater use of water during a warm spell, or the possibility of another fire in the same general area, can reduce the availability of water needed for proper firefighting.

Adverse Weather

Adverse weather plays additional roles in large-loss fires. In addition to the effects of extreme temperatures on the water supply, strong winds are often responsible for the spread of fire to adjoining property. Moreover, the firefighter who must perform his duties in either extreme cold or extreme heat not only feels the effects upon his own body, but is handicapped in his physical movements by snow or ice, strong winds, or heavy rains. Also, response time is increased by these conditions.

Sprinkler Failure

Where sprinkler systems have been installed, owners tend to feel that the hazard of fire has been eliminated. In the main, their complacency is justified; fire sprinklers have an enviable record. In 96 percent of the cases where sprinklers have been activated, the fire has been controlled.

In 4 percent of the cases, sprinkler failures have occurred for the following reasons:

1. Poor maintenance
   a. Sprinkler shut down
   b. Sprinkler tank dry
   c. Wet systems permitted to freeze
   d. Controlling valves on branch lines closed

Table 2-5 Operating Status of Buildings When Large-Loss Fires Occurred

STATUS	PERIOD	PERCENT OF TOTAL
Closed temporarily	Closed for night, holiday, weekend, or a period of less than one week	46.7%
In operation	Open for business as usual	41.9%
Closed for service	Not open to general public or employees, but attended by maintenance or service crews, inventory personnel, or other staff	5.7%
Closed, long term	Closed for the season, vacant, or under construction	5.7%
		100.0%

Source: Fire Journal, July 1972, copyrighted by the National Fire Protection Association. Reprinted by permission.

2. Poor housekeeping
a. Stock piles too high, blocking spray
b. Obstructions to proper operation, such as tables, benches, and racks that block access to burning stock

Incendiary

Currently, 27 percent of large-loss fires are listed as having been started maliciously. A percentage of those listed as "cause unknown" may also have been incendiary. This high figure indicates that the deliberately set fire often becomes a large-loss fire. The overall percentage of incendiary fires for all building fires is only 6.5 percent.

Most large-loss fires occur when the building is closed during normal nonwork periods, as shown in Table 2-5.

It is readily apparent how the small fire in a vacated building can spread. Of greater interest is the percentage of fires that occur in occupied buildings. Both figures point up the need for early detection systems and a means to control, or extinguish, the small fire by built-in extinguishing systems, such as automatic sprinklers.

Other Factors

Local geographic and weather conditions can affect local high-incidence peak periods. California, for example, experiences some of its worst fires when the hot-day "Santa Ana" type of wind is prevalent. A rash of forest fires usually occurs when humidity is low and a hot dry spell has been protracted. National figures, however, indicate the worst season of the year for fire appears to be the quarter from October through December when heating plants are turned on. The period of January through March is second in number of fires. Especially in January and February when heating systems are being pushed for very long cold spells, a large tabulation of heating fires in encountered. Another factor in the increase of alarms in these periods is confinement to the indoors because of weather conditions. The peak danger period for fire is between midnight and 2 a.m., with 91 percent of fire deaths and 90 percent of all large-loss fires occurring in the 12-hour period from 8 p.m. to 8 a.m., according to recent statistics.

The increased monetary loss due to fire over the years is cause for concern. Part of this loss can be accounted for by inflation. This, however, is not the sole answer. With the great strides in technological knowledge, new materials and processes have flowed into our way of life; some of them are not yet fully tested and are without proper safeguards for safety and fire. Also, a concentration of property values has increased exposure hazards and taxed available firefighting facilities. Population growth has speeded the deterioration of existing buildings, resulting in increasing fire loss.

On the positive side, statistics indicate that the ratio of loss to total property value has decreased. Even though the nation's population has soared above 200 million, the fire fatalities have stayed near the 12,000 per year figure for the past two decades. This fact can be accounted for in part by (1) improved firefighting techniques through better training and education of the firefighter; and (2) better and more versatile equipment.

SUMMARY

No organization can function properly without some form of record keeping: Statistics are needed to plan steps toward achievement of the organization goals and objectives. This requirement of course exists in the fire department organization. Fire statistics show what is burning, the causes and numbers of fires, and the loss of life and property in the various occupancy classifications. From these statistics, trends, causes, and possible solutions can be derived.

To be meaningful, the statistics must be reliable, a condition which predicates a uniform, standardized, and accurate reporting system. Only then can the department take proper control steps to achieve its objectives of the prevention and control of fire.

Information on the fire itself should include:

1. Area of origin

2. Equipment involved

3. Form of heat ignition

4. Type of material ignited

5. Form of material ignited

6. Commission or omission of an act by one or more individuals

To evaluate fire-fighting techniques, other information must be included, showing:

1. Fire units responding

2. Extent of fire

3. Method of extinguishment

4. Damage estimate

5. Weather conditions

6. Injuries or deaths

The large-loss fires, those responsible for 95 percent of the total dollar loss from fire, can be attributed to one or several of the factors listed as contributing to the large loss.

DISCUSSION TOPICS

1. What information on the fire itself must be included if a fire report is to prove meaningful and useful?

2. List other specific information that would be useful to the fire department and that should be included in the fire report.

3. What sources does the NFPA use in compiling national statistics on fires and fire losses?

4. Why should the dwelling fire be of particular concern to the firefighter?

5. List major causes of the large-loss fire.

RESEARCH PROJECTS

1. Gather statistics from a local fire department on fires and fire losses. Compare these data with national statistics and evaluate the department's method of reporting.

2. Request fire statistics from another municipality in another state; compare them with national statistics and evaluate the municipality's method of reporting.

3. Obtain blank forms used for fire reporting from other municipalities; then evaluate and compare them.

4. Using fire statistics as reported by a municipal fire department, outline areas of concern and control steps that might be taken.

5. A manufacturer is seeking a location for a new plant. Drawing on current statistics, outline concerns of the fire department that should be considered in the executive's planning.

FURTHER READINGS

American Insurance Association, special-interest bulletins, nos. 25, 37, 43, 44, 87, 95, 107, 109, 140, 151, 152, 187, 190, 257, 265, 296.

National Fire Protection Association, Bulletin No. 901, *Coding System for Fire Reporting*, 1971.

Tryon, George H. (ed.), *Fire Protection Handbook*, sec. 1, National Fire Protection Association, Boston, 1969.

Fire statistics indicate the scope of fire loss in the United States not only in property loss, but also in loss of life. The cost of fire has many facets. The main consideration, of course, is the cost that is experienced by the persons suffering the fire. This cost may be personal injury or, in many cases, loss of life. Property destroyed may vary from total loss of a building and its contents to minor property damage. The property loss may be treasured antiques or heirlooms, or irreplaceable business equipment. A fire in a business establishment can well mean loss of wages and loss of business.

In addition, there are indirect factors affecting many who are not directly concerned with the fire itself. Secondary victims may experience loss of work time or of personal business because of their inability to obtain the product of the company that had the fire. Other offshoot fire costs are fire insurance premiums, loss of tax revenue on the fire building, building costs attributable to fire code requirements, and of course the costs of the fire department and of private fire protection systems.

Consequently, many agencies and organizations are active in fire prevention and control research, development, and dissemination of information in both the public and the private sectors of the economy.

FEDERAL AGENCIES

The federal government has long been concerned with fire prevention and control measures through the media of the existing

FIRE PROTECTION AGENCIES

federal departments that have specific regulatory powers over certain designated areas.

Since the late 1920s and early 1930s, the period of the Great Depression, the federal government has taken a more extensive and active role in the country's social and economic welfare. Apparently, the federal government will also have to act more decisively and more extensively in the field of national fire prevention and control.

A step toward that end was the passage of the Fire Research and Safety Act of 1968 (see Appendix D). Funding of this act was not authorized until fiscal 1971. Funding facilitated the activation of programs aimed toward the realization of the broad goals of the legislation. The act provides for:

1. Investigation of causes, frequency, severity, and related factors of building fires

2. Research on fire causes

3. Development of better methods of fire prevention and fire control leading to a reduction of death, injury, and property damage

4. Education of the public to fire hazards and safety

5. Education and training of the fire service

6. Reference service on all aspects of fire safety

7. Projects demonstrating the new approaches

This legislation also provided for the establishment of a national commission on fire prevention and control, with its twenty members to be appointed by the President of the United States. President Nixon named the members in 1971, then charged them with the responsibility of submitting a study on the nation's fire problems. Information for their study was obtained in part in five national hearings, with testimony given by selected witnesses. Priority areas for study were (1) Prevention of fires; (2) early detection and warning of fires; and (3) fire suppression. The report of the study with findings and recommendations was given to the President and the U.S. Congress in July 1973 (see Appendix E).

Department of Commerce

The Fire Research and Safety Act as such falls under the jurisdiction of the National Bureau of Standards in the Department of Commerce. The National Bureau of Standards has long concerned itself with fire protection engineering in the establishment of standards for building structural components and the fire characteristics of various materials and erections used in construction. Standards for interior furnishings and clothing are also established to guard the public against fire hazards in these areas. The Flammable Fabrics Act of 1953 came out of the need for control of new materials in the clothing industry. Publications relative to work in the area of fire protection are available from the Superintendent of Documents, U.S. Government Printing Office, Washington 20402.

Department of Agriculture

The Department of Agriculture, while vitally concerned with the farm life of America, expresses its concern with fire safety by issuing pamphlets on fire prevention measures.

The United States Forest Service, within the Department, is responsible for 186.5 million acres of national forest and grassland spread over the United States. The Forest Service maintains its own division of fire control. Forest-fire research is carried out at nine experimental stations and a forest products laboratory. Control techniques and equipment have been developed that serve well the peculiar problems of forest firefighting. In addition to the use of aircraft in firefighting, the Forest Service developed "thicker" water that retards water runoff and has capabilities of absorbing more units of heat. "Thicker" water is now finding a place in conventional firefighting. Other chemicals have been developed that can be added to water and dropped on the fire scene from aircraft. The Forest Service's quarterly publication, *Fire Control Notes,* is available from the Superintendent of Documents in Washington.

Department of Interior

The National Park Service, in the Department of the Interior, is responsible for fire safety in the national park lands and buildings. It operates either through its own fire-fighting forces or through its supervisory control, by means of fire codes and regulations, in those areas where direct firefighting is the responsibility of the leaseholder of park territory.

The Bureau of Mines is concerned with mine safety. Mine-safety research at the Bureau's Pittsburgh research center has produced much information on explosions and fire that is pertinent to firefighting, although the fire service is not directly involved with mine fires and safety. Masks, devices important to the fire service, have been developed and tested by the Bureau of Mines.

Department of Labor

The Labor Department is directly concerned with occupational safety. Fire prevention inspection and training thus become part of its programs on safety standards under the Bureau of Labor Standards and the Office of Occupational Safety. Working with an arm of the Department of Labor are the Department of Transportation and the Interstate Commerce Commission. The Department of Transportation establishes standards, in the form of Transportation Department regulations, for the shipment of hazardous materials by truck, rail, water, air, or pipeline (see Chapter 12).

Department of Defense

All branches of the armed forces assume responsibility for fire protection and prevention in their services. Research and development in fire-fighting science and training in fire-fighting techniques are carried out.

Federal Fire Council

The federal government itself has property loss and loss of life due to fire. In a recent year, government buildings suffered about 37,000 fires, a $328 million property loss, 1,375 injuries, and 425 deaths.

The Federal Fire Council (Figure 3-1) was organized in 1930 on an informal basis. It has since grown to a full-time, officially recognized agency charged with the responsibility of disseminating fire-safety knowledge to federal government agencies. The council was recently transferred from General Services to the National Bureau of Standards, which administers the Fire Research and Safety Act.

Figure 3-1   The Federal Fire Council.

The President by Executive Order #7397
Authorizes Administrator
General Services Administration
Commissioner, Public Buildings Service
Chairman

Federal Fire Council
(250 members from 53 agencies)

EXECUTIVE COMMITTEE

Design Standards
Develops standards for fire protection and prevention in buildings.

Education and Training
Distributes educational materials—promotes training programs.

Fire Loss Experience
Collects, analyzes, and publishes loss data.

Fire Prevention
Establishes recommended practices and procedures.

Field Activities
Develops plans and projects to be implemented in the field. Aids establishment of association with field personnel by chartering field federal fire councils and certification of fire-safety groups in federal safety councils.

Program
Plans and arranges meetings.

Protection of Records
Develops standards for protection of files, films, and photographs.

Research and Technology
Develops understanding of fire technology and interprets fire research findings.

Systems and Equipment
Advises and develops standards for fixed and portable fire protection equipment.

The members of the council are such officers or employees of the various departments and establishments of the federal government that have been designated by their respective agency because of his or her responsibility, interest, or experience in fire safety.

From the past can be gained knowledge to assist in successfully shaping the future, and federal agencies long have recognized that fire-safety measures have to be integral with operations if the agencies are to fulfill their obligation to the public, the Congress, and the President—the obligation to perform their functions in the most efficient and economical manner.

Better coordination of fire-safety measures is expected to result. The council's three primary objectives are:

To reduce fire loss in the federal government

To assist in establishing fire-safety programs in all agencies

To encourage research in techniques of fire control and prevention

STATE AGENCIES

State Fire Marshal

Forty-four of the fifty states delegate responsibility of fire protection to a state fire marshal's office. As each state possesses autonomy and sovereign rights, the functions of the state fire marshal can vary from state to state. Broadly speaking, the marshal's tasks are as follows:

1. *Investigation of suspicious fires.* In many instances the state fire marshal is charged with the responsibility of complete investigation of a suspicious fire, turning only the prosecution over to the state's attorney. He has been granted uniquely broad powers in the right of entry for inspection and investigative purposes and the right to subpoena all records and all persons believed to have knowledge of the fire.

2. *Compilation of state fire-loss statistics.* These figures are used by the National Fire Protection Association in its published data on national fire losses.

3. *Other functions.* Some state fire marshals, if not all, are also responsible for:

Fire prevention education

Fire service training

Regulation of fireworks, flammable liquids, liquefied petroleum gas, and other hazardous materials

Regulation of fire protection systems, fire-alarm systems, and fire-extinguishing systems

Inspection of specified occupancies

A model fire marshal law detailing the duties of the state fire marshal has been recommended by the Fire Marshals' Association of North America. This model law offers guidelines to those states not having a state fire marshal's office and to those states whose marshals do not have clearly stated duties and legal rights.

State Forestry Departments

The state forestry department is responsible for fire protection and prevention for state lands, and, in some instances, county-controlled forest, brush, and grasslands. Thus, its responsibilities are similiar to those of the U.S. Forest Service for the national forests.

Office Civilian Defense (now Defense Civil Preparedness Agency—DCPA)

The Civil Defense Corps was activated during World War II as a home guard in defense against enemy attack. After the war, the Civil Defense remained active, mainly in alerting private citizens to the need for air raid fallout shelters as a necessary precaution against nuclear attack. Public shelters also were sought and stockpiled with supplies for persons caught away from home in a nuclear blast.

Today, public concern over a nuclear holocaust has subsided, yet the Civil Defense is still active. It now plays an important role in the natural or man-caused disaster, such as flood, tornado, or large fire. Much emergency equipment is in the inventory of the Civil Defense—some equipment has been donated, some is federal government surplus, and some has been bought on a federal matching-fund basis.

Civil Defense volunteers have been trained by police and fire training schools and are a definite aid to the local police and the fire service in an emergency. Even the names of many state civil defense units have been amended to be more relevant to the current need. For example, one such corps is called the Office of Emergency Service.

Education and Training

Many states offer training and education programs for the fire service; some prepare and sell training manuals and booklets on related topics. Many community colleges in many states are now offering a certificate or two-year degree program in fire science. A handful of colleges offer four-year degree programs tailored to the needs of the fire service.

The need for an educational program on a college level was pointed up during a conference at the Johnson Research Center in Racine, Wisconsin, in 1966. Ten of the nation's leaders in fire science met to "isolate and define" problems facing the fire service in the next decade. One of the twelve resulting statements focused on education, affirming that "Professional status begins with education." This was known as the Wingspread Conference.

Many of the nation's colleges have accepted this challenge and now offer fire science courses. The intent and purpose of these course offerings may be summed up in the following statement from a community college catalog.

> This curriculum is designed for students who desire to pursue a career in fire fighting. Graduates are prepared to secure employment in municipal fire departments, industrial fire departments, insurance companies, private agencies and other agencies dealing with fire safety adjustment. Opportunities are also found in equipment sales. Courses are instructed by persons familiar with the area of fire science and experts in their specialized areas.

A typical listing of course offerings includes:

Introduction to Fire Protection

Fire Protection

Hazardous Materials

Introduction to Fire Suppression

Fire Service Hydraulics

Introduction to Fire Prevention

Water System Analysis

Fire Supervision and Community Relations

Fire Department Administration

Technical Science

Materials of Construction

(The student is referred to *Guidelines for Fire Service Education Programs,* listed in Further Readings at the end of Chapter 4 for additional information on college programs.)

COUNTY AGENCIES

The individual county government's role in fire-safety techniques varies considerably also. Some counties have county fire marshals analogous to the state fire marshal. Other counties offer complete fire protection to the county by means of a county fire department. Notable examples are Dade County, Florida, and Los Angeles County, California. In areas not provided with municipal or county fire-suppression means, fire districts have been established within the county to provide fire protection.

THE PRIVATE ECONOMIC SECTOR AND FIRE PROTECTION SERVICES

Industrial Fire Departments

Large industrial plants and those having particular hazards may have an organized fire company or brigade made up of trained company personnel. This corps is a first line of defense to hold the fire in check or, if possible, to extinguish it before arrival of the fire company having jurisdiction. The plant brigade will have the type and amount of equipment deemed necessary to the hazards, perhaps including a completely equipped fire engine.

Privately Owned and Operated Fire Protection Services

In addition to industry-operated fire brigades, in some instances fire protection is being rendered to the public by a privately owned, profit-oriented fire department. Insurance companies were the forerunners in developing the privately owned fire department, operating companies to service their insured. Although the insurance companies no longer provide fire-fighting services, there are still privately owned departments. It goes without saying that these companies must be extremely cost-conscious, and yet provide service equivalent to that offered by the publicly owned department. A notable example is the Scottsdale, Arizona, fire department.

International Association of Firefighters

The International Association of Firefighters is a union affiliate of the AFL-CIO. It is the official representative of firefighters throughout the United States. In addition to its concern for the wage structure and working conditions of the firefighter, it is active in training and other educational endeavors that relate to mutual concerns in the fire service. The I.A.F. also publishes a monthly magazine containing many training and educational articles.

International Association of Fire Chiefs

The International Association of Fire Chiefs has over 7,000 members who are active fire chiefs or actively engaged in some phase of fire service. It is currently supporting greater professionalism in the fire service through college-level fire science courses. It supports its own Fire Administration Institute. Many of the members collaborate with the NFPA in establishing fire service standards.

An exchange of information among its members is conducted by means of seminars, conferences, a yearly convention, and a newsletter.

National Fire Protection Association

The NFPA is a nonprofit organization dating back to 1896. Its stated objectives are to promote and improve methods of fire protection and prevention by obtaining and circulating relevant information. It is dedicated to the task of establishing safeguards against loss of life and property. It seeks attainment of this goal by a diversified program.

Technical Standards. NFPA standards offer recommendations for fire protection and correction of hazards in a wide range of occupations and processes. These standards are highly technical in nature. Many of them have been incorporated in fire prevention codes and legislation. Although obtainable in individual pamphlet form, these pamphlets are published in a ten-volume set known as *The National Fire Codes.* Individual volume titles are:

1. *Flammable Liquids*
2. *Gases*

3. *Combustible Solids, Dusts, and Explosives*

4. *Building Codes and Facilities*

5. *Electrical*

6. *Sprinklers, Fire Pumps, and Water Tanks*

7. *Alarm and Special Extinguishing Systems*

8. *Portable and Manual Fire Control Equipment*

9. *Occupancy Standards and Process Hazards*

10. *Transportation*

The association also publishes three magazines slanted toward different branches of the fire service field. *Fire Journal,* published six times a year, is distributed only to association members. It covers technical developments in fire prevention and fire protection and also contains articles on major or unusual fires. *Fire Command* (formerly called *Firemen*) is a monthly whose articles and other content are directed to the fire-fighting service. *Fire Technology,* issued quarterly, is the publication of the Society of Fire Protection Engineers (SFPE), professional branch of the NFPA. It is directed toward the field of fire engineering. The SFPE is truly a professional society, and it has done much work toward creating an awareness of the work of the fire protection engineers and the need for them.

Others publications, in booklet, leaflet, poster, and pamphlet form, are designed to stimulate the knowledge of fire-safety practices in the general public, the property owner, the firefighter, and other agencies concerned with fire safety (see Figure 3-2).

Other units within the NFPA are the Fire Marshals Association of North America, the Electrical Section, the Railroad Section, and the Industrial Fire Protection Section.

Underwriters' Laboratories

In 1894 the National Board of Fire Underwriters created the Underwriters' Laboratories to test and examine devices, systems, and materials to determine their relation to life, fire, and casualty hazards. (See Figure 3-3.) The organization maintains testing laboratories and centers in various state locations and in twenty-five foreign countries. The main office and testing laboratory is in

Figure 3-2 Typical Occupancy Fire Record reports published by the National Fire Protection Association giving information on fire causes and contributing factors, with case histories of actual fires.

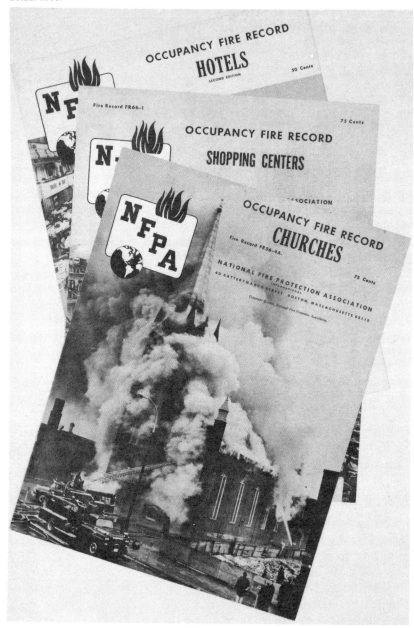

Figure 3-3   Underwriters' Laboratories.

The various types of new work engineering and Follow-Up Services, and other types of services provided by UL, together with the general method of product identification, where applicable, are described in the accompanying tabulation.

TYPE OF SERVICE	APPLICABLE TO	PRODUCT IDENTIFICATION
Listing	Those products which have been evaluated with respect to all reasonably foreseeable hazards to life and property, and where such hazards have been safeguarded to an acceptable degree.	*Listing Mark*—May appear in various forms as authorized by UL. (The following forms have been authorized.)
Classification	Those products which have been evaluated with respect to specific hazards only, or with respect to performance under specified conditions.	*Classification Mark* — May appear in various forms as authorized by UL, but includes a statement such as "Classified by Underwriters' Laboratories, Inc. With Respect to (Nature of Hazard) Only." The mark may also include an indication of the product's performance under a rating system designed for the particular product category.
Recognition	Those products which have been evaluated only for use as components of end-product equipment Listed, Classified, or Certified by UL.	*Recognized Markings*—Consists of the manufacturer's identification and catalog number, model number, or other product designation as specified for the particular product. Except in very rare cases, the use of the UL name in any form on the product is prohibited.
Certificate	Products comprising field-installed system at a specific location, or to specific quantities of certain products where it is impractical to apply the Listing Mark or Classification Marking to the individual product.	None. UL authorizes manufacturer to issue written certificate to customers or installers, based on UL's examination and/or tests.
Inspection	Products or systems involved in situations where the availability and capability of UL inspectors throughout the U. S., and in 30 foreign countries, can assist local authorities having jurisdiction, or can provide inspection service to an industry group on a contract basis.	None. A report covering the inspection is issued to the contracting sponsor.
Fact–Finding and Research	Projects conducted on contract basis for manufacturers, trade association, governmental agencies, and others in the interest of public safety. 　(1) *Fact-Finding Investigations* — develop product or system information and data for use by applicant in seeking recognition in or amendment of a nationally-recognized installation code or standard. 　(2) *Research* — develops basic information, properties, and characteristics of materials, products and systems as related to safety to life or property—generally in the area of standards development.	None. A report is issued to contracting sponsor. Reports generally cover a class rather than a specific product.

Chicago. Products that meet the internationally recognized standards developed by Underwriters' Laboratories are permitted to carry the UL approved label and to be listed in annually published lists of products proven acceptable. The product is thereafter subject to periodic checkups. The Underwriters' Laboratories maintains six departments that handle specific fields of testing:

Burglary Protection and Signaling

Casualty and Chemical Hazards

Electrical

Fire Protection

Heating, Air Conditioning, and Refrigeration

Marine

All testing and examination must meet safety standards developed and published by Underwriters' Laboratories. All the departments do some new work investigation that is relevant to the firefighter. The laboratories also work in the field of public information and education, distributing product lists of tested products and publishing fire-safety brochures among other activities.

A quarterly magazine and motion pictures explaining the laboratories' operations are available to interested persons. In addition, UL offers radio and television safety messages for general airing.

INSURANCE ASSOCIATIONS

Insurance associations have done much to promulgate fire-safety practices.

American Insurance Association

The A.I.A. is a nonprofit organization composed of the National Board of Fire Underwriters, the Association of Casualty and Surety Companies, and the American Insurance Association. It is probably best known as the publisher of the special-interest bulletins previously issued by the National Board of Underwriters. Until October 1,

1971, the American Insurance Association provided a municipal survey service. This service has been transferred to the Insurance Services Office.

Insurance Services Office

The Insurance Services Office (I.S.O.) is a consolidation of five national insurance industry service groups to provide insurers with improved statistical, actuarial, and research services. Thirteen lines of insurance, including dwelling and commercial fire, are serviced. Many insurance fire-rating bureaus have also been incorporated into the I.S.O. structure.

The I.S.O. now administers the Municipal Survey Service, which surveys and reports on the fire defenses of 465 large United States cities. It uses a standard grading schedule, first established in 1916 and periodically revised. (See Chapter 4 and Appendix C.)

In 1973 the I.S.O. published a new grading schedule to be used in classifying the fire defenses and physical conditions of a municipality. The new schedule places greater emphasis on water supply and fire department features than did the old schedule. Other important changes also have been made: Required fire flows are now based on fire potential rather than population, and the recommended number of engine and ladder companies is based on these established fire-flow requirements and travel distance rather than on a population formula. (Appendix C contains excerpts from the new grading schedule.)

Insurance Rating Bureaus

Each insurance rating bureau is a private, nonprofit organization established to determine a fair fire insurance rate for the state in which it is incorporated. Its name usually includes the state name, but the organization is financed solely through subscription of the fire insurance companies doing business in the state. It is subject to state examination and licensing.

Municipalities with less than 25,000 population have their fire defenses surveyed by the rating bureau. The grading schedule used by the A.I.A. and the Municipal Survey Service is the standard used.

American Mutual Insurance Alliance

The American Mutual Insurance Alliance (A.M.I.A.) is a national association of most of the large mutual insurance companies. In the main, it serves its members by providing technical know-how through engineering, legislative research, and educational dissemination. Service to the general public includes Operation EDITH (Exit Drills In The Home).

Factory Mutual Engineering Corporation

The Factory Mutual is known for the inspection work and recommendations made by its staff of fire prevention engineers in its insured properties. Standards have been established for industrial-plant fire protection pertinent to the building construction and occupancy. The *Factory Mutual Record* is a bimonthly publication of articles and editorials in the area of property conservation and risk management. Factory Mutual also operates a research corporation for the study of fire, explosives, and related hazards. This is a two-part research program—basic and applied. Basic research has the objective of securing information pertaining to the initial phases of fire, its detection, and its growth pattern. Applied research is concerned with work such as the inhibition of flames, explosions, and detonations; new suppression agents or systems; and ignition and flammability of materials.

Factory Insurance Association

The Factory Insurance Association is composed of capital-stock fire insurance companies that also offer engineering services and periodic inspections of the insured property. A fire-safety laboratory is operated for the examination and operational display of fire protection equipment.

The Factory Insurance Association publishes a bimonthly magazine, the *FIA Sentinel.* Other special brochures and booklets of a technical nature on fire prevention and fire protection may be obtained. Speakers are available for trade associations and fire protection organizations.

SUMMARY

Many agencies, other than the fire service, are concerned with fire and fire losses. They are active in both the public and the private sectors of the economy. Their operations include:

1. Testing of materials

2. Recommendations of standards for processes and materials

3. Recommendations of standards for fire defenses of municipalities

4. Occupational safety

5. Fire-control practices

6. Education and training

7. Fire investigation

8. Codes and ordinances

DISCUSSION TOPICS

1. List some direct and indirect costs of fire.

2. What important benefits has the fire department gained through the work of the Bureau of Mines?

3. Discuss important functions of the state fire marshal.

4. Discuss the evolution of the Civil Defense from World War II to its present status.

5. What was the "Wingspread Conference"?

6. List some magazines available to the firefighter that are pointed toward fire protection, fire prevention, or both.

RESEARCH PROJECTS

1. Select an agency in the private sector of the economy that is active in fire protection; research its activities.

2. Select an agency in the public sector active in fire protection; research its role in fire safety.

3. Outline the role the Fire Safety Act (Appendix D) can play in fire protection in the United States.

4. Detail the procedure followed in the testing of a particular process or material by a recognized agency.

FURTHER READINGS

The following agencies are actively engaged in fire protection. Many offer pamphlets and reports. A complete listing of their services may be obtained by writing directly to the agency.

American Insurance Association, 85 John St., New York 10038. National building code, fire-resistive ratings, fire prevention codes.

American Mutual Insurance Alliance, 20 N. Wacker Drive, Chicago 60606.

American Society for Testing and Materials, 1916 Race St., Philadelphia, Pa. 19103. ASTM standards in building codes.

Factory Insurance Association, 85 Woodland St., Hartford, Conn. 06102.

Factory Mutual System, Factory Mutual Engineering and Research, 1151 Boston-Providence Turnpike, Norwood, Mass. 02062.

Federal Fire Council, 19th and F Sts., N.W., Washington 20405.

Fire Marshals Association of North America, 60 Batterymarch St., Boston, Mass. 02110.

Insurance Services Office, Municipal Survey Service, 85 John St., New York 10038. (Now responsible for municipal surveys formerly conducted by American Insurance Association.)

International Association of Fire Chiefs, 1725 K St., N.W., Suite 1108, Washington 20006.

International Association of Firefighters, 905 16th St., N.W., Suite 404, Washington 20006.

National Fire Protection Association, 470 Atlantic Ave., Boston, Mass. 02110. Standards and fire codes.

Underwriters' Laboratories, Inc., 207 East Ohio St., Chicago 60611. Standards.

U.S. Department of Agriculture, Forest Service, Division of Forest Fire Research, Washington 20234.

U.S. Department of Commerce, National Bureau of Standards, Washington 20234. Reports on building materials and structures, commercial standards and simplified practice recommendations.

U.S. Department of Interior, Bureau of Mines, Pittsburgh Research Center, 4800 Forbes Ave., Pittsburgh, Pa. 15213.

Wingspread Conference on Fire Service Administration, Education and Research, The Johnson Foundation, Racine, Wisconsin.

Under our federal form of government, individual states maintain certain states' rights and hold sovereign power in many spheres. Under this sovereign power, all other local government units, such as cities, villages, towns, and counties, owe their existence to the state government. They have been granted lifeblood by legislative action and charter of the state government.

The charter will usually contain authorization for the community to form its own fire protection service. In some communities this service may become a fire bureau within a department such as a department of public safety. In the larger municipality, the fire protection service will usually be organized specifically as the fire department, an arm of the municipal government.

The objectives of a fire department have been stated as follows:

1. To prevent fires

2. To prevent loss of life and property when fire does occur

3. To confine a fire to the place of origin

4. To extinguish the fire

To obtain these objectives, a twofold attack is launched: fire prevention and fire suppression. In the fulfillment of its task, the fire department is aided by other city departments—water, building, city planning, and police. Together, they compose the fire defenses of the municipality.

FIRE DEPARTMENT ORGANIZATION & ADMINISTRATION

Further, the fire protection activities, like other government responsibilities, may overlap outside the central municipality in some form of mutual protection agreement. The state and federal governments are also concerned with fire prevention and extinguishment. The state compiles statewide data on insurance fire losses, investigates arson and incendiary fires, and inspects factories and places of assembly. Also, it may have a coordinated program for handling disasters and civil defense. Some states maintain state-level fire departments (in California, the Division of Forestry). The federal government is concerned with fire protection in farm and rural areas, forest fires, marine fire safety, and the safety of government buildings.

The type of fire department organization depends upon the size of the city, the scope of activities planned, and the form of local government.

The department's authority should be vested in one head rather than in a board of control. Ideally, this person is solely responsible for administrative leadership—a unified command direct and continuous in all departmental activities. It is this person's responsibility to coordinate activities with, and to cooperate with, other departments for effective and economic service to the public. In the department itself, responsibility and authority may be delegated to subordinate staff officers who report to, and are responsible to, the administrative leader—a line-and-staff form of organization.

The head of the fire department may be titled Fire Commissioner, Fire Chief, or Chief Engineer. He is usually appointed by the mayor or city manager, normally with the advice and consent of the city council. The appointment will be for an indefinite term; he can be removed at any time for sufficient cause.

Other ranks in the department where employment is full time will normally be under civil service with the probable, but not universal, exception of office clerks and typists.

The fire department head will exercise direction and supervision over the employees of his department. "He shall prescribe the rules and regulations, practices and procedures for the department." It is his duty to report to the city head and the city council or fire board the activities and fiscal needs of his department as they affect personnel, equipment, apparatus, and quarters.

In essence, the department is scalar, or hierarchical in command, a line type of organization. The chain of command flows downward from the head through the subordinate chiefs to the individual company commanders and from the company commanders to the firefighters. Authority and responsibility are delegated to subordinates who are given definite spans of control and supervision, but final responsibility rests with the department head.

The organizational goals of a fire department are not unique to the fire service alone. The fire chief or commissioner is a business manager, the top executive of a public service organization. If the stated goals of the fire service are to be achieved, principles of sound business management that have served in the private sectors of the economy can be utilized.

In attaining management objectives, certain functions or processes must be considered. These processes involve planning, organizing, directing, and controlling.

PLANNING

No business can succeed without sound business planning, nor can the fire department. Basic planning will:

Establish objectives

Set policy

Weigh cost factors

Determine organizational responsibilities

Sound fire department planning will also consider present and future needs, including those relating to manpower, buildings and equipment, and techniques in the strategy and tactics of firefighting.

ORGANIZING

The organizational table is a method of work assignment and delegation of authority and responsibility to the ends of goal achievement. Management principles have long been established in the business world. They are based on the theory of *chain of command.* As described earlier, chain-of-command theory is utilized in most fire departments.

Three principles of management are basic to an effective chain of command:

1. *Unity of command.* While the top executive is ultimately responsible for the actions of all subordinates, each subordinate or worker will receive orders and directions from but one superior.

2. *Span of control.* Each subordinate with delegated authority and responsibility will have a restricted number of persons under his or her command. Generally a span of control of one to seven persons results in the most satisfactory interpersonal relationships.

3. *Authority and responsibility.* Where responsibility has been delegated to subordinates, the subordinate must be given sufficient authority to fulfill the responsible role effectively.

The most basic form of business organization employing these principles is of the simple line type. Other names for this type of organization are "military" and "hierarchical." The military must expect and get instant obedience to orders and commands out in the field. To assure instant obedience from a well-disciplined force, responsibility for its actions must be vested in the person who issues the orders. The chief executive is in direct control and responsible for all decisions, great or small.

The fire department head out in the field must also expect and get instant action and compliance with his orders and commands. Time, tide, and fire wait for no one. This parallel need with the military for well-disciplined subordinates, trained to instant action, lends to the fire service paramilitary characteristics. The line type of organization exists in the basic fire department and in small business firms.

Table 4-1 shows the line type of organization in three fields.

As the business expands or the fire department grows, administrative responsibilities increase and it becomes virtually impossible for one single executive to supervise all areas. The chief executive may then employ administrative specialists in the various areas to submit recommendations. The department head still maintains ultimate responsibility, but a staff of special advisers has been added to the line organization. This line-and-staff organization is common in large business organizations and city fire departments. A model organization chart of this type appears in Table 4-2.

Table 4-1 Comparative Line Organizations

ARMY	FIRE DEPARTMENT	BUSINESS FIRM
Major General	Chief	President
Brigadier General	Deputy Chief	Vice President
Colonel	Assistant Deputy Chief	Assistant Vice President
Lieutenant Colonel	Platoon Commander	Divisional Manager
Major	District Chief	Plant Manager
Captain	Battalion Chief	Department Manager
First Lieutenant	Company Officer—Captain	Section Manager
Second Lieutenant	Company Officer—Lieutenant	Supervisors
Soldiers	Firefighters	Workers

Each position in the chain of command is responsible for a specific part of the work that must be done to accomplish the objective of the organization under the direct control and guidance of those above in the chain of command. Many of the positions shown in the table may be eliminated in the smaller business or fire department. Such organizations are simple to set up; the line of authority is clearly established; and the top executive exercises direct control.

Among other disadvantages in this type of organization, the upper-level managers tend to become bogged down with minor details. Conversely, these same persons may lack the technical knowledge or specialized skills required of many managerial positions.

The administrative functions served in the line-and-staff fire department organization include:

Budgets and payroll

Personnel records

Planning

Training

Equipment and supplies purchasing

Equipment maintenance

Communications

Fire control and suppression

Fire prevention

Fire investigation

Public relations

Table 4-2   A Model Fire Department Line-and-Staff Organization

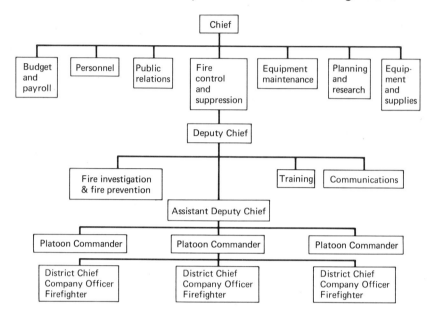

A line type of organization with staff-delegated functions added. In some instances these staff functions are performed by personnel outside the uniformed fire service. Rank held in staff functions is dependent on the size of the department.

Each of these administrative functions may be under the supervision of a different manager, but unless the department is very large, one manager probably will handle several allied functions.

Most of these functions support the function of fire control and suppression. Yet, a stated goal of the fire department, and the most important one, is the *prevention* of fire. But fires do happen. So, by necessity, most manpower, equipment, and monies are allocated to fire control and suppression.

The administrative functions may be summarized thus:

1. *Budgets and payroll,* including payroll, general accounting finances, and budget allocations.

2. *Personnel records,* consisting of personnel history and records, including medical services; and manpower allocations, requirements, and recruitment.

These two functions are best performed by workers who, although not included in the uniformed force, are under the direction and guidance of a staff officer. In the smaller departments, these functions may in the main be handled by a central municipal bureau.

3. *Planning and research,* involving both the future needs in manpower, equipment and buildings, and research on fire-fighting techniques and the development of new equipment.

4. *Training,* providing not only basic training programs for the probationary firefighter but also frequent refresher or specialized training programs for firefighters at the training center. In addition, a daily company-level training session should be held. It may consist of both drill evaluations and classroom lectures and discussions. Critiques of fires may also be conducted in the classroom period.

5. *Equipment and supplies,* including *(a)* normal office supplies, records, automotive supplies, and automotive and building repair and maintenance supplies; and *(b)* equipment and supplies needed in fire control that must be maintained at needed levels. This function may fall under a central purchasing bureau of the municipality.

6. *Equipment maintenance,* with the proper repair and maintenance of automotive equipment and tools necessary for fire control a must. The fire company cannot function without equipment and tools in good repair. This function, particularly the maintenance of automotive apparatus, may be carried out by a municipal repair and maintenance division.

7. *Communications,* including the main function of receiving alarms and dispatching equipment to the fire scene. Maintenance and repair of alarm systems and interdepartmental communication systems may be further functions.

8. *Public relations,* with the fire department maintaining its own public relations officer. In addition to training programs, fire inspection, and actual fire duty, firefighters still have time to become involved in community functions. Their activities may involve appearances before community groups, talks to school children, and conducting tours at the fire station.

The local fire company goes out into its fire response area and visits the various occupancies. This is known as a prefire planning inspection. Its purpose is to acquaint the firefighter with the occupancy and its potential hazards and to plan what tactical action would be taken in the event of a fire. A preplanning inspection is an excellent means to promote better public relations. A sincere interest in others' property and a desire to lessen a person's possible fire damage should help not only to control fire loss, but also to promote goodwill.

The growing trend, especially in small departments, is for the local company also to take on fire-safety or fire prevention inspection work. This activity differs somewhat from the preplanning inspection in that code violations are specifically sought out and cited. In the preplanning inspection, minor violations are brought to the attention of the owner for immediate correction, and apparent major violations are referred to the fire inspection officer for evaluation and corrective action. In many instances, the company will make both types of inspection at the same time. The apparent punitive action of the fire prevention inspection thus tends to cancel out the good will of the preplanning inspection.

Many other social and economic problems in the community can be aided by the time and effort of the local firefighters—supervision of sports activities for the young, aid to older citizens, help in fund-raising drives, for example. These are a few of the means of building better community relations between the firefighter and the tax-paying citizen.

9. *Fire investigation,* involving the determination of the cause of fire when it is not readily apparent and the responsibility for making arson investigations of all suspicious fires, will be a function of a fire investigation unit acting as a separate unit or as an arm of the fire prevention section. A photography laboratory that is responsible for photographs of arson scenes, large fires, accidents, and personnel photos may be included in this unit.

10. *Fire prevention,* involving fire prevention inspections of all industries and buildings other than residential structures. (On request, they may be inspected also.)

Violations of fire ordinances are noted and recommendations are made on the inspection trip. A follow-up inspection is made to see whether corrections of previously cited violations have been completed. Periodic reinspections follow.

A program for public education and safety is also an ongoing activity.

Records of all inspections and activities are maintained.

Because of the scope and responsibility of fire prevention activities, the officer in charge should hold high rank in the staff hierarchy.

(Chapter 14 provides additional information on fire prevention and investigation.)

11. *Fire control and suppression,* carrying responsibility for tactical application of fire extinguishment and rescue operations in the field and for the saving of life and property. This is the line function toward which most staff functions are pointed.

The basic fire department line unit is the fire company—the ladder-truck unit or the engine pumper. The company officer is directly responsible for the subordinates under him. In the unity or chain of command, all orders and directions from above will filter down through this officer to the company firefighter. All requests and other communications from the firefighter will pass through him to the hierarchal level above.

The engine company may respond alone on a fire call described as an auto, brush, grass, fence, or rubbish fire. On an alarm of any other fire, a basic attack team, consisting of two engine companies, a hook-and-ladder company, and a chief officer, should respond. At the chief's discretion, further help will be called for if needed. His call will bring more of the department's own equipment or aid from adjoining departments and additional chief officers.

Rescue operations may include an ambulance service in addition to the numerous nonfire emergencies to which the fire department responds.

DIRECTING

Having established an organizational table showing clear lines of responsibility and delegating the commensurate authority needed to achieve the stated goals, the department must then ensure that proper action will be taken toward the ends desired. This is the task of direction, of guiding the organization by leadership and command. The department leader will normally execute this task by the continual issuance of orders to those under his command.

The directing officer may follow a laissez faire policy or use an autocratic-dictatorial approach—leaders differ in personality and method. But the most successful director will also consider the individual personalities of those he is directing. By following basic principles of human relationships, he will stimulate their pride of accomplishment and their desire to achieve the goals outlined.

CONTROLLING

The best-laid plans of mice and men go oft astray. Control is based upon the premise that organization objectives are not easy to obtain and that a constant reevaluation of performance weighed against objectives must be made. Corrective measures may need to be taken in the steps toward achieving the objectives.

Factors that affect fire department performance are always present. Personnel may change, new hazards may be encountered, new techniques in firefighting may be developed. These elements must be under constant evaluation, as they may necessitate a change in planning.

The goals of the fire department are high and not easy to attain. Proper leadership and coordination of effort and control measures help to keep these goals as worthy objectives and to bring meaning to fire department organization.

THE COMPANY COMMANDER

The local fire company unit is the backbone of the fire department. As the leader of the fire company, the company commander is in one of the most responsible positions of command leadership and the one most difficult to fulfill.

He is responsible for the initial attack on the fire scene and, as the low man on the hierarchical ladder of command, he must serve as the buffer between labor (the firefighter) and management (the deputy chiefs and the chief above him). All orders and commands filter down to him, and it is essentially he who sees that the orders are carried out, at the same time considering the needs and capabilities of the firefighters beneath him. The importance of his role can be seen in the following responsibilities:

1. Welding a heterogeneous group of individuals into a functioning fire company

2. Upholding the dignity and integrity of the individual

3. Maintaining a fire station in a clean and healthful condition

4. Maintaining fire apparatus and equipment in proper condition

5. Ensuring proper and efficient usage of equipment by a program of company in-service training

6. Rendering service to the tax-paying public with dispatch and in a professional manner when called upon

7. Maintaining records, reports, and orders

THE FIREFIGHTER

The effort toward professionalism in the fire service also has brought on a reevaluation of the standards for job qualification. Standards, including physical, medical, and educational requirements, vary to a certain degree in various municipalities and sections of the country. However, guidelines are available, such as those offered by the International Association of Firefighters in its "Recommended Standards for Firefighters."

The following general guidelines provide a model that is more or less standard for most departments.

Admission Requirements

The prospective applicant must be a citizen, between nineteen and thirty years of age, and of good moral character. Conviction for a crime that carried a prison sentence is usually grounds for rejection. The educational minimum requirement is high school graduation or the equivalent.

Medical and physical requirements must be more stringent than for less strenuous occupations. The candidate must possess normal vision, normal hearing, and sufficient muscular development. There must be a proper relationship between height, weight, and age. The condition of heart, lungs, and respiratory system is very important and may be cause for rejection by the fire service, although no handicap would be apparent in other endeavors.

The physical examination will test general agility and muscular development necessary for the raising and climbing of ladders, maneuvering of hose lines, and the general physical activities required of a firefighter.

A written examination is also a part of the entrance procurement examination. This examination is general in nature, testing for general aptitudes rather than for specific firematic knowledge.

A personal interview completes the testing program. This interview enables the examining board to gauge the prospective candidate's appearance, demeanor, attitudes, and ability to communicate.

Both the personal interview and the medical examination may reveal personality quirks or medical symptoms that would limit the applicant's ability to get along with and live together with fellow firefighters. This ability is necessary because of the nature of firefighting work and the working hours and conditions.

Having successfully passed the entrance requirements, the applicant is placed on an eligibility list in rank order with the grade mark attained. Vacancies in the department are filled from the top of this eligibility list downward.

Probation and Training

Applicants who have been called to fill a vacancy will undergo a probationary period of six months to a year. Their actions will determine how well they can assimilate the training given to them in this probationary period, and of equal importance, how well they can adapt to fire department discipline. During the probationary period, they may be discharged for valid reason without a hearing. After completion of the probationary period and final acceptance, they can be dismissed only by the filing of charges by the chief officer with the civil board, which must then act on these charges. If, after an unbiased hearing, they are discharged, firefighters may still resort to the civil courts for judicial review.

In the large department, the firefighter will receive basic training at the department's training academy. Recruit training for the small department may be done under state supervision at certified training facilities by certified instructors using a state-recommended curriculum.

Here the firefighters will be given a thorough grounding in firefighting principles, practices, and procedures. Schooling consists of both classroom and drill area procedures. After completing the basic training, they will report to a fire company. Training at the company level will continue on a daily basis throughout their time in the fire department.

Career Opportunities and Security

The chances for promotion are good, provided they are worthy, conscientious, and capable of absorbing the added knowledge and responsibilities required at the next grade level.

They have chosen the profession of firefighter for several reasons:

1. They seek a vocation that will provide the thrill and excitement and the physical activity that they need.

2. They are genuinely interested in serving humanity.

3. They seek job security and old-age security. As firefighters, they have the job security through the civil service protection of dismissal only through valid proof of cause. They have fair old-age security through a state or city pension code relating to a firemen's annuity and benefit fund. A portion of their salary goes toward this pension fund. An equal amount is contributed by the city. They have no Social Security benefits.

Pension systems, injury pay, voluntary and compulsory retirement ages may all vary from municipality to municipality. However, the provisions of one fire department may serve as a possible model or guideline: Firefighters can retire, usually with half-pay, after reaching a minimal age of fifty and working a minimum of 20 years in the fire service. If they elect to continue working after the minimum retirement age, 2 percent per year is added to the pension until they reach compulsory retirement at sixty to sixty-three years of age. If they are forced to retire because of a duty injury, they receive three-quarter pay until the compulsory retirement age. In the event of death either through duty injuries or off-duty injury or illness, a surviving spouse and minor children receive a pension.

MANPOWER

The new grading schedule for municipal fire protection as published by the Insurance Services Office in 1973 establishes a standard of six members on duty for each required engine and each required ladder truck company. Chief aides and ambulance personnel may be counted as part of this staffing, dependent upon their availability for actual fire-fighting operations. Some credit is also given for call and volunteer members and off-shift response.

It is difficult for most cities to meet these standards, and more and more, the statistics of our daily life work against their being able to do so.

Salaries account for 91 to 95 percent of a fire department budget. The remainder covers equipment, repairs, supplies, and administrative overhead. The cost of fire protection, like everything else, has been on the increase during recent years. General inflation, shorter working hours, and increased hourly wages have affected the fire service. The fire service finds itself in the position where it must lower costs and at the same time increase "protection."

Several factors are working toward this goal. The all-metal hydraulic-raise aerial ladder in the main has replaced the spring-assisted manually raised aerial that required possibly four people to raise fully and place into position. One person can position the hydraulic raise and do it in less time. Power tools are assisting in opening floors and roofs. Power charges for ventilation and forcible entry are being experimented with. Aluminum metal ladders, which can be raised by fewer people, are replacing wooden ladders.

Engine companies now operate not only with preconnected rubber booster lines and preconnected 1½-inch lines but with preconnected 2½-inch lines and 2½-inch lines preconnected into fixed master-stream devices. Thus instant water can be applied at the fire scene; in using these fixed master streams, one man can operate the turret nozzle.

The five-person engine and the six-person truck have given way to the four-person engine and the five-person truck. The realities of rising living costs and the tax burden on the homeowner may well bring about further manpower reductions. If such is the case—and many small communities cannot now afford the luxury of four-person engines and five-person trucks—another possible answer will be task-force response. Rather than two or three pieces of apparatus converging on a fire scene from different directions at differing times, however slight, the response will be as a unified force. Three or four companies can be housed in one station, all responding together. Housed together, the task force should result in savings by the elimination of single-company quarters. Responding together, task-force efforts can be coordinated and directed toward the task that needs doing first. And in a joint effort, the two lines deemed necessary at the average fire may be laid with fewer people and in less time than with two pumping engines working separately.

The grading schedule for municipal fire protection also uses established criteria to determine the number of pumping engines and aerial ladder trucks thought necessary for the proper protection of a municipality with a given population and fire potential. Other factors that can affect fire protection are also considered.

The number of engine and ladder companies needed under the new grading schedule of the I.S.O. are based upon travel distance to the fire potential and the required fire flow. Fire flow is defined in the grading schedule (see Appendix C) as follows:

Required fire flow. The required fire flow is the rate of flow needed for firefighting purposes to confine a major fire to the buildings within a block or other group complex. The determination of this flow depends upon the size, construction, occupancy, and exposure of buildings within and surrounding the block or complex; consideration may be given to automatic sprinkler protection.

Again, economic considerations dictate that the municipality quite often is short of attaining these specified goals. However, the trend today is for the municipality to amend building codes to permit construction of buildings of greatly increased height as a compensation for high land values. The fire department of the community is thus given the added burden of having a larger aerial ladder and the know-how to attack a fire in the high-rise building.

SUMMARY

Principles of management and established organizational fundamentals apply to the fire department as well as to modern business. Fire department objectives cannot be achieved without them. A model organization chart can be used as a guideline in evaluating the individual municipality needs.

The fire department is a service organization, an arm of the civil government operating under state control. Guidelines have been established to determine manpower and equipment requirements. Economic considerations of the municipality sometimes negate the fulfillment of these requirements.

The responsibility then lies with the fire department to establish sound practices and procedures, to devise new methods and uses of equipment so that it can realize progress toward its objectives.

DISCUSSION TOPICS

1. Discuss the operational functions of a simple line organization.

2. Compare a line-and-staff operation with a line organization.

3. List the main functions performed in any type of organization and the main goal of each function.

4. Why is the role of the fire company commander so important?

5. What are some basic standards for firefighter applicants?

RESEARCH PROJECTS

1. Obtain and evaluate the organizational chart of a local fire department.

2. Using the appropriate table from the grading schedule in Appendix C, determine the number of pumpers and aerial trucks needed within travel distance of established fire flows of 2,000 gpm and 11,000 gpm.

3. Determine the manpower requirements for the pumpers and aerial trucks recommended in the preceding projects.

4. Set up a model organization chart, based on needed fire equipment, for the municipalities having established fire flows of 2,000 gpm and 11,000 gpm respectively.

FURTHER READINGS

Dale, Ernest, *Management: Theory and Practice,* McGraw-Hill, New York, 1965.

Dale, Ernest, *Organization,* American Management Association, New York, 1967.

Faureau, Donald F., *Fire Service Management,* Reuben H. Donnelley, Inc., New York, 1969

Nolting, Orin, *Municipal Fire Administration,* International City Managers Association, Chicago, 1967.

Odiorne, George S., *Management by Objectives,* Pitman, New York, 1965.

Fire, or combustion, can be defined as rapid oxidation with the generation of heat and light. All matter oxidizes. Iron rusts, paper yellows and then darkens and becomes brittle. These examples are slow processes that go on for months or perhaps years. In fire or combustion, however, the oxygen in the air-oxidizing process acts very swiftly on matter.

A study of the combustion of fire cannot be made without some knowledge of the natural sciences. Natural scientists have long been concerned with the phenomena of matter, its actions, reactions, and interactions. They have formulated theories, laws, and models to test their theories. As new knowledge is gained, it may necessitate change in some theory and also in the model. Much of this new knowledge has direct application to the field of fire technology.

What, then, is fire? What element can cause a substance to be consumed beyond recognition? In point of fact, what is "substance" itself?

THE STRUCTURE OF MATTER

Kinetic Molecular Theory

Natural scientists tell us that substance, or matter, is anything that occupies space and possesses weight or mass. It exists in one of three stages. It can be either a solid, a liquid, or a gas. It is capable of passing through the three stages of solid–liquid–gas because of the action of various processes. Matter can undergo other changes also, as evidenced by the process of cooking or the act of burning.

Matter, then, is capable of undergoing both physical change and chemical change. A physical change is characterized by a change in

ELEMENTS
OF FIRE

form; glass may be formed into a window pane or a drinking vessel. Both are composed of the same kind of matter. Another type of physical change occurs when liquid becomes solid or a solid liquefies. Water as ice is matter of a kind that has undergone a physical change. A physical change is usually a transient one.

A chemical change is characterized by more extensive and permanent changes that can be reversed only by certain laboratory processes.

The kinetic molecular theory, relating to matter, has been formulated. This theory states:

1. Matter is made up of small, discrete particles called molecules. All molecules of a specific matter are identical and differ from the molecules of other matter.

2. Molecules have a field force that exerts an attraction on other molecules and binds them together.

3. Molecules are in rapid, ceaseless motion. As they collide with one another and with the walls of the container in which they are confined, they exert a pressure. As the temperature rises, their motion becomes more rapid. Kinetic energy increases with this temperature rise and the pressure becomes greater.

Matter appears to be a solid mass when in a so-called solid state. Yet the kinetic molecular theory tells us that the individual molecules composing the matter are in swift, unceasing motion. In the solid state, the kinetic energy is at low ebb. There is a freedom of molecular movement, but the freedom of each molecule is restricted by the low energy level and the binding field forces of its neighboring molecules.

The natural scientists also tell us that energy, the ability to perform work, can be neither created nor destroyed; it can only be transformed. This is known as the conservation of energy. Heat is a form of energy.

Most matter can exist in a solid, liquid, and gaseous state. As heat in the form of energy is applied to the solid, the law of the conservation of energy will operate thus: The solid will absorb the heat energy and this increased energy level will enable the molecules to move about more rapidly. The size of a molecule is estimated to be 10^{-8} centimeters in diameter. (A molecule, then, is the equivalent of one-billionth of a centimeter.) The increased movement will open the distance between molecules to about the size of the molecules themselves. One molecule can escape the field force of its neighbor, but it immediately comes under the momentary influence of other molecules. Although energy levels are not high enough for a complete breakaway, there is a constant motion and restricted freedom. This freedom permits the solid matter to become liquid. This state is characterized by the ability of the liquid to assume the shape of the container in which it is placed, but with a coherence of its own.

As additional energy is applied in the form of heat, molecular motion will increase. The vacant spaces between the molecules will be many times the size of the molecules. Their rapid, random movement will exert a pressure against the sides of the container as they strike its walls. The field force bond is still present, but the swift movement negates the bonding force of the attraction. As the energy level of the gas rises, pressure against the restricting vessel will increase. This is the gaseous state. *The expanding gas will fill the container, exerting a pressure if confined,* or escaping to the atmosphere if the container is open.

Thus far, we have seen that matter is made of particles called molecules. We are also aware that there are many types of matter, each type possessing a molecular structure peculiar to its own kind and differing from that of other matter.

Molecules do change. Where a chemical change takes place, such as through cooking or burning, the molecular structure becomes that of a different matter. In theory, nature has a building block smaller than the molecule. This smaller building block is the atom, the smallest unit of matter, and yet a single entity. When the atoms in a molecule are rearranged, they form a different type of molecule. This chemical change process produces a new matter with its own identifiable molecule. This process of formation of a new matter can take place by:

1. Decomposition: The existing molecule of complex structure breaks down into simple-structure molecules.

2. Synthesis: The existing molecule is added to, and builds up to, a more complex molecule.

3. Rearrangement: Two or more molecules of different matter exchange parts and, in the rearrangement, form a different matter.

Science tells us that at the present time, there are only 103 known specific particles. They are composed of one or more atoms of identical kind, and are available for the formation of matter. These particles, called elements, are listed in Table 5-1.

Atomic Theory

In 1808, John Dalton stated the fundamental tenets of the atomic theory:

1. The number of atoms is limited.

2. All atoms of a kind are identical and indivisible and cannot be altered. They differ from the atoms of any other matter.

3. Molecules are combinations of atoms, and all molecules of one kind have identical atomic structure.

4. Chemical change is caused by the formation of new molecules of a different substance.

From this limited number of basic building-block atoms, all known substances are formed. It can be mathematically shown that the possible kinds of matter that can be created are numberless. It is known, however, that the ten most common kinds of atom make up more than 99 percent of the earth's crust, seawater, and atmosphere. These ten atoms, and the percentages of their occurrence in our universe, are as follows:

Oxygen	49.5%	Sodium	2.6%
Silicon	25.8%	Potassium	2.4%
Aluminum	7.5%	Magnesium	1.9%
Iron	4.7%	Hydrogen	0.9%
Calcium	3.4%	Titanium	0.6%

All others 0.7%

A summary may help to crystallize the previous statements.

A molecule may consist of only one atom or of several atoms, which may be either identical or of different types. Those molecules that are composed of one atom or of only one kind of atom are known as *elements.* As there are at present only 103 known atoms, we can know only 103 elements. Some of these elementary substances are well-known gases and solids and are either metallic or nonmetallic.

Some examples of elementary substances are:

GASES	SOLIDS
Helium	Copper
Oxygen	Gold
Hydrogen	Sulfur
Nitrogen	Uranium
Chlorine	Carbon

Those substances that are composed of more than one kind of atom are called *compounds.* A compound is a new substance formed by the combination of one or more elements, an element and a compound, or two or more compounds. It is in the compound form that innumerable kinds of matter, each with its own identifiable molecule, can be formed.

Some examples of common compounds are:

COMPOUNDS	ELEMENTS
Water (H_2O)	Hydrogen + oxygen
Table salt (NaCl)	Sodium + chlorine
Sulfuric Acid (H_2SO_4)	Hydrogen + sulfur + oxygen
Ammonia ($2NH_3$)	Nitrogen + hydrogen
Baking soda ($NaHCO_3$)	Sodium + hydrogen + carbon + oxygen

A compound differs from a mixture. A mixture is also a combination of elements, of an element and compounds, or of compounds. But, in the mixture, the substances can be identified and they maintain their own identity. Sand and sugar combined are a mixture, but each substance can be identified and separated without the process of chemical change.

Electrons, Protons, and Neutrons

Present theory states that the structure of the atom consists of three fundamental particles—electrons, protons, and neutrons. The proton has a positive charge, the electron has a negative charge, and the neutron carries no charge. The number of protons and the number of electrons in the atom are equal. Each known atom has a

Table 5-1 Chemical Elements*

ATOMIC NUMBER	ELEMENT	SYMBOL	ATOMIC WEIGHT
1	Hydrogen	H	1.008
2	Helium	He	4.003
3	Lithium	Li	6.940
4	Beryllium	Be	9.02
5	Boron	B	10.82
6	Carbon	C	12.010
7	Nitrogen	N	14.008
8	Oxygen	O	16.000
9	Fluorine	F	19.00
10	Neon	Ne	20.183
11	Sodium	Na	22.997
12	Magnesium	Mg	24.32
13	Aluminum	Al	26.97
14	Silicon	Si	28.06
15	Phosphorus	P	30.98
16	Sulfur	S	32.006
17	Chlorine	Cl	35.457
18	Argon	A	39.944
19	Potassium	K	39.096
20	Calcium	Ca	40.08
21	Scandium	Sc	45.10
22	Titanium	Ti	47.90
23	Vanadium	V	50.95
24	Chromium	Cr	52.01
25	Manganese	Mn	54.93
26	Iron	Fe	55.85
27	Cobalt	Co	58.94
28	Nickel	Ni	58.69
29	Copper	Cu	63.54
30	Zinc	Zn	65.38
31	Gallium	Ga	69.72
32	Germanium	Ge	72.60
33	Arsenic	As	74.91
34	Selenium	Se	78.96
35	Bromine	Br	79.916
36	Krypton	Kr	83.7
37	Rubidium	Rb	85.48
38	Strontium	Sr	87.63
39	Yttrium	Y	88.92
40	Zirconium	Zr	91.22
41	Niobium	Nb	92.91
42	Molybdenum	Mo	95.95
43	Technetium	Tc	98. +
44	Ruthenium	Ru	101.7
45	Rhodium	Rh	102.91

ATOMIC NUMBER	ELEMENT	SYMBOL	ATOMIC WEIGHT
46	Palladium	Pd	106.7
47	Silver	Ag	107.880
48	Cadmium	Cd	112.41
49	Indium	In	114.76
50	Tin	Sn	118.70
51	Antimony	Sb	121.76
52	Tellurium	Te	127.61
53	Iodine	I	126.92
54	Xenon	Xe	131.3
55	Cesium	Cs	132.91
56	Barium	Ba	137.36
57-71	The rare earths—comprising the very rare Lanthanide series		
72	Hafnium	Hf	178.6
73	Tantalum	Ta	180.88
74	Tungsten (Wolfram)	W	183.92
75	Rhenium	Re	186.31
76	Osmium	Os	190.2
77	Iridium	Ir	193.1
78	Platinum	Pt	195.23
79	Gold	Au	197.2
80	Mercury	Hg	200.61
81	Thallium	Tl	204.39
82	Lead	Pb	207.21
83	Bismuth	Bi	209.00
84	Polonium	Po	210 ±
85	Astatine	At	210 ±
86	Radon	Rn	222
87	Francium	Fr	223 ±
88	Radium	Ra	226.05
89	Actinium	Ac	227
90	Thorium	Th	232.12
91	Protoactinium	Pa	231
92	Uranium	U	238.07
93	Neptunium	Np	239 ±
94	Plutonium	Pu	239 ±
95	Americium	Am	242 ±
96	Curium	Cu	243 ±
97	Berkelium	Bk	247 ±
98	Californium	Cf	249 ±
99	Einsteinium	Es	254 ±
100	Fermium	Fm	253 ±
101	Mendelevium	Md	256 ±
102	Nobelium	No	254 ±
103	Lawrencium	Lw	257 ±

*Elements 61 and 93 through 103 are synthetically prepared, and their atomic weights are only inferred.

different number of protons in its nucleus. This, in essence, means that there can be an atom with one proton and one electron (hydrogen) or with as many as 103 electrons and 103 protons (lawrencium). The number of protons in the atom is indicated by the atomic number we have assigned to each element.

The electron has, for all practical purposes, no mass or weight. The proton and the neutron have a mass established as unity, or one.

The number of neutrons in the atom has no relationship to the number of electrons and protons. The sum of the number of protons and neutrons gives the atomic weight of the atom. Conversely, the atomic weight less the atomic number (the number of protons) shows the number of neutrons in the atom. In some cases, atoms of the same atomic number have different atomic weights. This disparity indicates that the two atoms have different numbers of neutrons in the nucleus. These variances are known as isotopes of the element. Some elements have as many as ten isotopes.

Chlorine, hydrogen, and carbon are simple examples of common elements that possess natural isotopes. Nuclear bombardment has produced some 750 isotopes not found in nature, all radioactive. There is at least one radioactive isotope for every element.

The protons and neutrons make up the nucleus of the atom. An earlier model of the atom had the negatively charged electrons circling the nucleus much as the planets and the earth circle the sun. The electrons were known as "planetary" electrons. The latest scientific model of an atom shows the electrons moving in a wavelike motion in an electronic cloud about the nucleus. (See Figure 5-1.) These electrons are arranged in energy levels, known as orbitals, depending upon the number of electrons in the atom. The closer the electron is to the nucleus, the less energy it contains.

The level closest to the nucleus of the atom has room for two electrons. An atom having more than two electrons will have the third electron in the second energy level. The second energy level has room for eight electrons. When eight electrons have filled this energy level, all remaining electrons must go to the third energy level. The number of electrons permitted in each successive energy level or group is determined by the formula $2N^2$. N represents the group level. In the case of the third group or level, this would be 2 times 3^2, or 18. However, this level, or any other, will not contain more than eight electrons until a successive level has been started.

Figure 5-1 Model of the electron cloud theory of motion. Although the electrons arrange themselves in energy levels about the nucleus of the atom and revolve around this nucleus, they travel in a spherical cloud rather than in a fixed orbit. The dots indicate various locations of the electron in its travel. The largest numbers of dots indicate the most frequent locations.

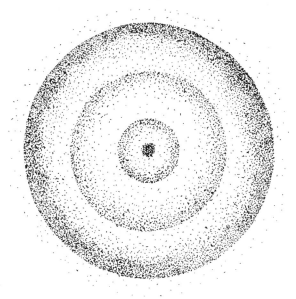

This simplified explanation suffices for us to remember that no outermost level or group will ever contain more than eight electrons. If a level does contain more, it is no longer the outermost level. Thus electrons appear to have an affinity for a grouping of eight.

Using this theory, the elements can be grouped in a periodic table from 1 to 8. The placement of an element within a group is governed by the number of electrons on the outer shell orbital. This periodic table indicates the chemical behavior of the element, there being great similarities between elements with the same number of electrons in the outer shell. (Figure 5-2 shows a simplified periodic table.)

Classification of Elements

The elements are identified by symbols—normally the symbol is the first letter of the element, usually with one other letter. For example, calcium has the symbol Ca; oxygen, O; and chlorine, Cl.

Figure 5-2 Simplified periodic table of the elements.

Group

IA							VIIIA
1							2
H	IIA	IIIA	IVA	VA	VIA	VIIA	He
3	4	5	6	7	8	9	10
Li	Be	B	C	N	O	F	Ne
11	12	13	14	15	16	17	18
Na	Mg	Al	Si	P	S	Cl	Ar
19	20	31	32	33	34	35	36
K	Ca	Ga	Ge	As	Se	Br	Kr
37	38	49	50	51	52	53	54
Rb	Sr	In	Sn	Sb	Te	I	Xe
55	56	81	82	83	84	85	86
Cs	Ba	Tl	Pb	Bi	Po	At	Rn
87	88						
Fr	Ra						

Some of the symbols of the elements are taken from their Latin names. To mention a few, K is the symbol for potassium (in Latin, *kalium*); Na represents sodium (*natrium*); and Pb stands for lead (*plumbum*).

As said earlier, the elements are either metals or nonmetals. Some of the nonmetal elements are oxygen, hydrogen, carbon, and nitrogen, the so-called organogens or components of all organic matter. Another group consists of the halogens—fluorine, chlorine, bromine, and iodine. The halogens are related and have similar properties. For example, each element in the halogen group has the same number of electrons in the outer shell. Other nonmetal elements are sulfur, phosphorus, boron, and silicon.

The nonmetal elements, in combination with hydrogen or hydrogen plus oxygen, form the so-called acids.

The metallic elements can be broken into the following subgroups:

1. Alkali metals—potassium, sodium, and lithium, as well as cesium, rubidium, and francium. All are in group 1 of the periodic table, having one electron in the outer shell.

2. The calcium group, including calcium and magnesium. (Barium, strontium, beryllium, and radium complete group 2, having two electrons in the outer shell.)

3. Active heavy metals (the iron group), consisting of tin, lead, iron, aluminum, nickel, cobalt, chromium, manganese, and zinc.

4. Less active heavy metals, namely, copper, arsenic, silver, platinum, gold, antimony, and mercury.

The metals, in combination with hydrogen or hydrogen plus oxygen, form a hydroxide of the metal known as a base. This base, in union with an acid, forms the salts.

Some elements, or atoms, spontaneously change themselves into different kinds of atom without going through the process of a chemical change. In this spontaneous change, a gaseous emanation is given off. When in this change process, the substance is undergoing nuclear change and is radioactive. These radioactive elements are constantly emitting energy. The energy emanated is in the form of alpha, beta, and gamma rays. Theory also states that these elements are in the process of disintegrating when the alpha or beta particles are emitted. The emitting atom becomes the atom of another element, and radioactive decay is said to have taken place. The length of time required for half the radioactive material to decay is known as its half-life. This half-life is important in the study and control of radioactive materials.

Theory and discovery of nuclear change, coupled with Einstein's theory that mass and energy are the same, opened up the new field of atomic energy.

Hydrocarbons

All living organisms have the elements of oxygen, hydrogen, and carbon as components. So important is carbon that the field of organic chemistry has been established as a study of this one element and its compounds. Hydrocarbon compounds, composed of hydrogen and carbon, make up a large category in the field of organic chemistry.

These hydrocarbon compounds, varying from a gas to a solid, are of particular importance to the firefighter. Also significant to him are both the chemical change that takes place through combustion, and the chemical change that occurs when elements or compounds are mixed in the presence of a liquid such as water.

A study of the elements of fire and the hazards to the firefighter will in the main focus on chemical change through combustion or the addition of water. A separate category of hazards to the firefighter includes those resulting from an element or compound with which he comes in contact by accident: a leaking chlorine tank, an exposed radiation hazard, a corrosive acid spill.

A substance known as a compound, being formed of more than one kind of atom, will be represented by a molecular formula that shows the kinds and numbers of atoms that are present in that compound.

For example, it is a well-known fact that water is composed of the atoms of hydrogen (H) and oxygen (O). The molecular formula of water, H_2O, tells us that there are two hydrogen atoms present for each oxygen atom in water:

$$2H + O = H_2O$$

Sulfur (S) plus two parts of oxygen (O) composes sulfur dioxide (represented in the formula SO_2). The formula for sulfuric acid is H_2SO_4. This tells us that two atoms of oxygen undergo a chemical change that results in the new substance of sulfuric acid. By formula,

$$2H + S + 4O = H_2SO_4$$

A return to the periodic table theory will throw further light on the formation of compounds. The electrons in the outer level or group are known as valence electrons. As previously stated, these electrons feel more comfortable or stable in a grouping of eight. In order to achieve this feeling of stability, the outer valence electron tends to be shared with electrons in an identical element or in a compound. This sharing of outer electrons is known as a covalent bond. Let us examine some examples of electron sharing or covalent bonds.

Chlorine atoms have seven electrons in the outer level.

$$:\overset{..}{\underset{..}{Cl}}\cdot \qquad \cdot\overset{..}{\underset{..}{Cl}}:$$

By sharing the one electron, the two chlorine atoms achieve stability.

$$:\overset{..}{\underset{..}{Cl}}:\overset{..}{\underset{..}{Cl}}:$$

Now consider an example of a compound. Hydrogen has one electron. Oxygen has six electrons in the outer shell. Thus,

$$H\cdot \qquad H \qquad \cdot\overset{.}{\underset{..}{O}}: \qquad = \qquad \begin{array}{c} H \\ \\ H \ \ :\overset{}{\underset{..}{O}}: \end{array}$$

The oxygen atom now has stability with eight electrons and the compound H_2O or water has resulted.

Sulfur and oxygen both have six valence electrons. In the formula SO_2, the two oxygen atoms share valence electrons with the sulfur atom. The two oxygen atoms thus achieve the stability of eight.

FUELS

Some of the elements, or compounds, are known as fuels. Fuels are those substances that will burn when subjected to a degree of heat peculiar to the particular element or compound. While some elements, such as magnesium, titanium, and sulfur, are fuels, the most common elements considered as fuels are hydrogen and carbon. Nearly all plant and animal life contains carbon, hydrogen, oxygen, and nitrogen. Oxygen is not a fuel but, as we shall see later, it contributes to and supports combustion. Nitrogen is not a fuel, nor will it support combustion.

Most compounds that are considered as fuels will be found to contain these organic elements of carbon, hydrogen, oxygen, and nitrogen. Wood, paper, and cotton all contain, as part of their molecular formulas, the compound cellulose.The molecular formula for the compound cellulose is $C_6H_{10}O_5$. Petroleum-flammable liquid compounds are known as hydrocarbons and contain the atoms of hydrogen and carbon in varying formulas. Gasoline can vary in structure from the formula C_5H_{12} to C_9H_{20}.

Flammable liquids and gases all contain elements of carbon and hydrogen. For example, ethane consists of C_2H_6; propane, of C_3H_8; and butane, of C_4H_{10}.

The covalent bond for ethane works out thus: Carbon has four valence electrons. Hydrogen has one valence electron.

$$
\begin{array}{ccccc}
 & H & & H & \\
 & \overset{x.}{} & & \overset{x.}{} & \\
 & x & x & & x \\
H & \cdot\; C & & C\; \cdot & H \\
 & & x & & \\
 & \underset{H}{.x} & & \underset{H}{x.} & \\
\end{array}
$$

(The cross represents carbon electrons; the dot, hydrogen electrons.)

The covalent bond can be shown in all hydrocarbon formulas.

It bears repeating that, although the known fuels are not limited to compounds of the atoms carbon and hydrogen, these two atoms are present in most fuels.

If a substance that is capable of burning, a fuel, is necessary for fire to take place, what then is fire? Further research into the natural sciences is necessary before our answer can be given.

THE ATMOSPHERE AND OXYGEN

The atmosphere—the air around us—is composed of oxygen (21 percent) and nitrogen (for all practical purposes, 79 percent). (There are approximately 1 percent argon and a 1 percent mixture of other gases considered too minute to catalog here.) In other words, the air we live in is approximately one-fifth oxygen and four-fifths nitrogen. In the presence of oxygen, all matter undergoes change. Again, paper tends to yellow with age, and iron is said to rust as it slowly forms iron oxide in a chemical change. Both materials can be said to be undergoing a process of oxidation. This slow oxidation process takes place at the temperature range that will also support life, or what is known as ambient temperature, the normal temperature of our environment. When temperatures exceed the normal or ambient temperature and continue to rise, the oxidation rate also increases. When materials such as paper or wood (fuels classified as combustible) unite with oxygen in a rapid oxidation process, the material is said to be burning, or in the process of combustion. The light, heat, and flame being given off are said to be fire. Fire, or combustion, then is defined as the rapid oxidation of a substance with the evolution of heat, light, and flame. If flame is not present, the fire can be said to be in a smouldering, or inactive, stage.

It can also be said that the combustible material has reached its ignition temperature. Ignition temperature is defined as the minimum temperature at which a solid, liquid, or gas initiates or sustains combustion. Most combustible materials, or fuels, must have access to oxygen to undergo the process of combustion. Normally, this oxygen is present in the atmosphere, or air, and without the presence of air, most combustible substances will not burn. There are exceptions, however, as noted later. These exceptions release enough oxygen when in a chemical reaction process so that combustion takes place without the presence of air.

Oxygen itself will not burn, but it is a contributor to, and a supporter of, combustion. The higher the oxygen content, the hotter will be the heat of the combustion and the faster will be the oxidation or burning process. Attesting to this is the combination of acetylene gas and pure oxygen into a flame of sufficiently intense heat to weld or cut steel. The acetylene gas is incapable of generating this amount of heat in a normal atmospheric condition, but needs the increased oxygen mixture to burn more vigorously and with more heat. A more tragic example of this is the rapid fire and the loss of the lives of three astronauts undergoing training at Cape Canaveral. An electrical short in an atmosphere of 100 percent oxygen caused a fire that consumed the interior of their spacecraft and killed them before they could release an escape hatch.

Again, for all practical purposes, combustibles must be accessible to air before combustion can take place. With a richer mixture of oxygen, combustion will be hotter and more rapid. Conversely, if the oxygen content is lowered, combustion will be slowed or less complete. If the oxygen content in the air is lowered from the normal 21 percent to 15 percent, combustion will be stopped completely or will not be initiated. From these facts it can be seen that the nitrogen in the air acts as a fire retardant or as an agent that slows down the rate of combustion.

Human beings, too, need oxygen for the oxidation process of living. Just as the process of combustion slows down to complete extinguishment when oxygen content is lowered, the process of human life slows down until life cannot be sustained when the oxygen content drops to only 6 to 10 percent.

THE FIRE TRIANGLE

In reexamining the rapid oxidation process known as combustion, we note that three factors appear to be necessary:

(1) A combustible material is needed in (2) the presence of oxygen or an oxidizing agent, with (3) a source of heat sufficient to increase temperature of the combustible material to its ignition temperature.

These three factors have been incorporated into the simplistic fire-triangle model *air—heat—fuel,* producing *fire.* This concept is shown in Figure 5-3.

Once combustion has been initiated, and given an ample supply of fuel and oxygen, the fire can become self-supporting. As the fuel burns, it creates more heat. The increase in heat raises more fuel to its ignition temperature. As the need for more oxygen arises to support the combustion, more oxygen is drawn into the fire zone. (In large conflagrations or raging forest fires where heat is intense and fuel abundant, a windstorm of air bringing more oxygen to the fire zone can be observed and felt.) The oxygen in turn increases the heat of burning and more fuel becomes involved. While oxidation is speeding up to the combustion stage, another process is occurring that aids in the process of combustion. A substance exposed to the action of heat will undergo a chemical decomposition process known as pyrolysis. As chemical decomposition takes place, the substance will give off vapors and gases which at certain temperatures can form flammable mixtures with the air.

This chain of reaction and interaction continues until either all the fuel has been consumed, all the oxygen used up, or heat has been dissipated so that the temperature of the fuel is lowered below its ignition temperature. This in essence states the fundamental method of fire extinguishment: Removal of one side of the fire triangle by:

Figure 5-3   The fire triangle—a simplistic model of combustion.

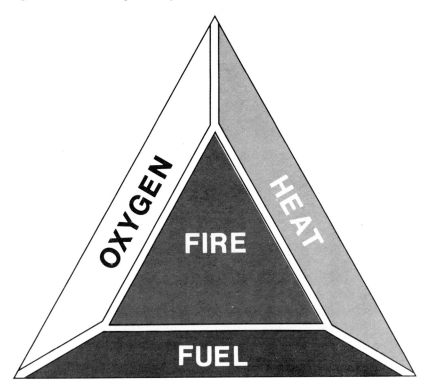

1.   Cooling, to reduce the temperature of the fuel to below its ignition temperature.

2.   Smothering, to prevent oxygen from reaching the fire by displacing the air with an inert gas, by sealing off within an inert blanket of foam, or by smothering the fire in some other way.

3.   Separation, by removing unburned fuel from the fire by physical or mechanical moving, or by using something, such as a water curtain, that blocks the passage of heat to the fuel.

THE TETRAHEDRON OF FIRE

More recent studies of the chemistry and physics of combustion have added another dimension to the simplistic fire triangle. A new model has now been theorized. With the development of dry chemical extinguishing agents and the halogenated hydrocarbon extinguishing systems, a question arose as to what exactly was causing extinguishment. Chemical extinguishment seemed to be more effective than that extinguishment effected through the normal process of cooling, smothering, or separating fuel from the source of heat. When the dry chemicals were first used, it was believed the formation of carbon dioxide as a product furnished the extinguishing power. It was soon determined, however, that the sodium bicarbonate base dry chemical had twice as much effect as an extinguishing agent as did an equal amount of carbon dioxide.

An examination of the chemical reaction that occurs in normal combustion gave credence to a new hypothesis of chemical extinguishment:

When a source of fuel, a source of heat, and oxygen combine, fire is the result. In the chemical reaction that takes place, there are many byproducts. Some of them are carbon monoxide (CO), carbon dioxide (CO_2), and sulfur dioxide (SO_2). More importance to these byproducts will be given in the sections on smoke and gases in Chapter 6. Other byproducts are free atoms of oxygen and hydrogen, known as radicals. These may form hydroxyl radicals with the symbol OH. Two of these OH radicals (2OH) may break down into H_2O and a free radical of O. This O radical feeds the chain reaction of combustion (discussed earlier) and contributes to the chain that greatly expands the fire.

It is now believed that the chemical extinguishing agent releases a chemical that "absorbs," or reacts with, the free hydroxyl radicals so that the oxygen is no longer free to contribute to the chain reaction. This chain has been broken. As an example, in the case of Halon 1301 extinguishant,[1] one of the chemicals released from the Halon 1301 is hydrogen bromide (HBr). The hydroxyl radical combines with the hydrogen bromide. The result is water and bromine. That is,

$$OH + HBr = H_2O + Br$$

The bromine radical then reacts with more fuel to remove more radicals from the fire—the chain reaction is broken and combustion

[1]Halon 1301 is one of the halogenated extinguishing agents containing two of the halogen series elements. Halon 1301 is a trademark.

stops. The fire triangle of cooling below ignition temperature, smothering by exclusion of oxygen, or separation of fuel from heat has played only a very minor part in the resultant extinguishment. The fire triangle now takes the form of a polyhedron composed of the original three sides of oxygen or oxidizing agent, fuel or reducing agent, heat or temperature, and a fourth side of uninhibited molecular chain reactions. This new model takes the form of a tetrahedron. (See Figure 5-4.) To our three methods of extinguish-

Figure 5-4   The fire tetrahedron model of combustion. The tetrahedron is a polyhedron with three sides and base. The sides and the base each represent a factor in this combustion model.

ment in the simplistic triangle, a fourth method can be summarized for the fourth side of the tetrahedron:

4. Suppression, by breaking up the formation of free radical compounds contributing to the spread of fire by the introduction of chemical extinguishment agents which react with the free radical atoms and inhibit the chain reaction.

There is, of course, an overlapping of extinguishment functions, and any given extinguishing agent may be reacting to more than one side of our new tetrahedron fire model. But more of this later.

SUMMARY

To understand why fire occurs and what the nature of fire itself is, one must have some understanding of basic chemistry. Matter may exist as a solid, a liquid, or a gas. The energy level is highest in the gaseous state, which expands the volume and exerts a greater pressure on the containing vessel. All matter is composed of elements which in turn are composed of atoms, which consist of fundamental particles of electrons, neutrons, and protons. There are more than 100 known elements from which all matter, however complex, is made.

In the formation of matter from the elements, there is a sharing of orbital electrons found on the outer shell of the element structure. This sharing is known as covalent bond, a search for a balance of eight electrons on the outer shell.

Some of the elements are fuels, the most common of which are hydrogen and carbon. They exist in combination in the hydrocarbon compounds. Hydrogen and carbon, along with oxygen and nitrogen, are found in nearly all plant and animal life. Most fuel compounds that burn will contain these organic elements. Oxygen itself will not burn, but it contributes to and supports combustion.

Combustion, or fire, is defined as the rapid oxidation of matter with the evolution of heat, light, and flame. For combustion to take place, there must be a combination of a fuel, a source of heat (all matter must be raised to what is known as its ignition temperature before it will burn), and the presence of air, or oxygen. This is the fire triangle.

It follows, then, that if a substance is cooled to a temperature below its ignition temperature, is denied access to air or oxygen (there are a few exceptions to the need for oxygen or air), or is separated from the source of heat, the combustion process will stop.

Latest research indicates that a chemical extinguishing agent such as Halon 1301 extinguishes by breaking up the chain reaction initiated by the combination of heat, fuel, and air.

A fourth method of extinguishment must now be included—suppression.

The simplistic fire triangle has now become a tetrahedron.

DISCUSSION TOPICS

1.   Discuss briefly the kinetic molecular theory.

2.   Why is the compound cellulose so important to the firefighter? What is its composition?

3.   Discuss the importance of hydrocarbons and the role they play in fire.

4.   Relate the three sides of the simplistic fire triangle to fire and the suppression of fire.

5.   Why was a fourth side added to the fire triangle? What is this fourth side?

RESEARCH PROJECTS

1.   Cite some experimental evidence that led to the kinetic molecular theory.

2.   List some reasons why a substance going from a solid or liquid to a gaseous state may be of concern to the firefighter. Give examples.

3.   Research the role of isotopes in nuclear energy.

4.   Describe three laboratory experiments that would serve to illustrate three methods of fire extinguishment. Explain what action took place.

5.   Research extinguishing agents, other than Halon 1301, that work on the principle of suppression.

FURTHER READINGS

Campbell, Richard D., *College Chemistry, A Survey,* Harcourt Brace Jovanovich, New York, 1968.

Garard, Ira D., *Invitation to Chemistry,* Doubleday, Garden City, N.Y., 1969.

Patton, A. R., *Science for the Non-Scientist,* Burgess, Minneapolis, Minn., 1962.

Pauling, Linus, *College Chemistry,* 3d ed., Freeman, San Francisco, 1964.

Porter, George, *Chemistry for the Modern World,* Barnes & Noble, New York, 1962.

Vaczek, Louis, *The Enjoyment of Chemistry,* Viking, New York, 1964.

The process of combustion is a chemical process. As combustible materials go through the oxidation process, a chemical action takes place. Materials that are reactive to air or oxygen are being converted into new substances of gases, water vapor, and an inert ash.

In order better to understand the process of combustion, it is again necessary to return to the chemist and the physicist for an explanation and a definition of the various factors that enter into the combustion process.

HEAT ENERGY

Recalling Chapter 5, we know that as the temperature of a substance rises, the motion of the molecules increases and becomes more rapid. Heat is a measurement of this molecular motion in a substance. In a solid, the molecular vibration is low and the substance remains a solid; as heat energy is transferred to the substance and as temperature increases, the molecules tend to spread further apart, and the solid becomes liquid. With a further increase in temperature, the molecular action becomes so rapid that the molecules, now in a gaseous state, leave the container that held them as a liquid. Once into the atmosphere, the molecules bounce off, and fill, the vacant spaces between the atmosphere's molecules. It can be said the gas has diffused into the atmosphere.

The source of heat energy that causes increased molecular activity and a physical change in the affected substance can come from the following sources:

1. Chemical energy through oxidation
2. Electrical energy sources
3. Mechanical heat energy sources
4. Nuclear heat energy

COMPOSITION OF FIRE

Chemical Oxidation

As explained earlier, chemical oxidation is the rapid oxidation that causes combustion. The numerous materials capable of rapid oxidation are known as combustibles. The most common of them contain large amounts of carbon and hydrogen.

Normally, sufficient heat for combustion is attained when combustible material absorbs heat from some adjacent material serving as an ignition source. But, in some cases, certain combustibles are capable of a self-generated increase in temperature up to the point where ignition can occur. This is known as spontaneous heating. While most organic substances do oxidize and release heat, this process is usually slow enough so that the induced heat is dissipated before combustion takes place. Spontaneous heating is so rapid that combustion occurs.

Electrical Heat Energy

Electrical energy can cause heat high enough to be the cause of fire through arcing, dielectric heating, induction heating, or through heat generated by resistance to the current flow. This last process may be intentional heating, as in the case of filaments or heating elements, or accidental, as when electrical "shorts" or overloading occur.

Static electricity is also responsible for an arcing effect when frictional electricity between a positively and a negatively charged body becomes great enough so that a spark is discharged between the two bodies. This spark may not be of sufficient duration or generate enough heat to ignite ordinary combustible materials, but may ignite flammable liquid vapors and gases.

Lightning has an action similar to that of static electricity. It occurs when a charge on a cloud arcs to the ground or to a cloud with an opposite charge. The magnitude of a lightning charge generates sufficient heat in many cases to ignite combustible materials. The high amperage and high voltage potential, although of short duration, can do much structural damage even though fire does not occur.

Mechanical Heat Energy

Mechanical heat energy can be caused by the heat of friction or the resistance to the motion of two bodies rubbing together. A slipping belt and a tire skidding on ice are good examples of heat generated through friction. Another source of mechanical heat energy is derived from the compression of gases. The temperature of the gas is increased when the gas is compressed. This can be simply demonstrated by the action of an air pump compressing air into an automobile tire or tube. As the pressure builds up, the tube valve and pump fitting gain heat that can be easily detected by the hand.

Nuclear Heat Energy

The release of very large quantities of energy from the nucleus of an atom is known as nuclear heat energy. One pound of the isotope of uranium, known as U^{235}, is capable of releasing the amount of energy obtainable from 5 million pounds of coal or the equivalent explosive force of 15,000 tons of TNT.

Nuclear heat energy can be released from the atom by two methods. Nuclear fission is the splitting of the nucleus of the atom. Nuclear fusion is the fusion of the nuclei of two atoms. Controlled release of the nuclear energy is finding a wide use in industrial technology, but concomitant with its use is a potential hazard.

Energy can be neither created nor destroyed. Matter can be neither created nor destroyed. These are the laws of the conservation of energy and the conservation of matter. In the reaction of one substance with another, there are a displacement of energy, a displacement of matter from the original substances, and the formulation of new substances. However, the total quantity of matter and the total quantity of energy will be the same.

For example, a combustible material burned in air will leave a residue of ash. The ash is considerably smaller than the original combustible matter. Air also has been consumed. And, in addition to the residue of ash, gases and smoke particles have been emitted into the atmosphere. The mass of the combustible matter and air is equal to the mass of the residual ash and emitted gas and smoke particles. The potential energy levels are also equal. Energy is being neither created nor destroyed, but only transferred.

Some of the potential energy of a substance undergoing change is in the form of heat. Those chemical reactions where heat energy is given up are known as exothermic reactions. Those newly formed substances which have absorbed energy in the form of heat are known as endothermic reactions. The first two letters in each term are the key identifying letters: *ex* signifies exit or loss of heat, and *en*, the entrance or addition of heat. (The two types of reaction are illustrated in Figure 6-1.)

A combustible substance that is burning is undergoing an exothermic reaction. Heat is being released from the substance.

Spontaneous heating is an example of an endothermic reaction. A substance is absorbing heat and its temperature is rising. The heat may continue to rise to the point where the substance too will be "on fire" and will then become involved in a heat-releasing exothermic reaction.

HEAT TRANSFER

A source of heat is necessary to raise the temperature of a combustible material to a point where combustion takes place. This heat source must be transferred to the combustible material by some means or method.

Heat is transferred from one substance to another by three methods: conduction, convection, and radiation.

Conduction

When two bodies are touching one another, or when each is touching the same solid, liquid, or gas, heat is conducted from one

Figure 6-1 Reaction in the endothermic-exothermic process (*a*) at the low-energy level, and (*b*) at the high-energy level. In an endothermic reaction, energy in the form of heat is added and the energy level of the substance is raised. In an exothermic reaction, energy in the form of heat is lost and the total energy of the substance falls.

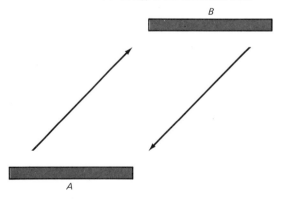

to the other through the medium of *conduction*. The greater the heat differential, the more rapid will be the rate of heat transfer. Also, some materials have a higher thermal conductivity rate, which permits a greater amount of heat to be conducted to the second body. All materials conduct heat, although some materials are classified as "insulation," or heat-retardant materials. These insulating materials do have a lower heat conductivity rate, but if heat is conducted through them faster than they can dissipate it, the second substance will take on heat.

The thickness of the medium between the two materials will also affect the transfer of heat.

Conduction of heat from one substance to another then depends on the temperature differential between them, the thermal conductivity of the two materials, and the thermal conductivity and the thickness of any materials separating them. Figure 6-2 shows examples of heat transfer by conduction in everyday cooking and in an uncontrolled fire.

Convection

When heat is transferred through a circulating medium rather than by direct contact, it is said to be transferred by *convection*. The medium can be either a gas or a liquid. The gas or liquid is first heated by conduction from another substance having a higher temperature. The heat is then transferred throughout the entire gas or liquid medium by convection. Where the medium contacts another substance of lesser temperature, some of the heat of the medium will be transferred to that other substance, again by conduction. It is through convection, with the medium being heated air, smoke, other fire gases, and some radiation, that an entire building may become heated to the point where it bursts into flame—"flashover" has occurred. (See Figure 6-3.)

Radiation

The sun is the supplier of all heat energy on earth. Radiant energy waves race through the atmosphere with a speed of light until they strike a substance where they are absorbed, reflected, or transmitted. These radiant energy waves radiate from the sun as infrared rays, visible light, and ultraviolet rays.

Figure 6-2   The transfer of heat by conduction. (a) An example of heat conduction in everyday living. The gas flame heats the frying pan by direct contact (conduction). The hot frying pan in turn cooks the food in the pan (again by conduction). (b) An example of heat conduction in an uncontrolled fire. The fire heats the wall separating the stored container from the fire. The heated wall, by conduction, heats the container and its contents until they also burst into flame.

(a)

The sun, then, is the classic example of heat transfer through *radiation.* But objects on earth that emit rays in wavelengths that vary from infrared to ultraviolet also emit radiant heat waves. Any hot object is sending out invisible infrared rays which tend to produce warmth or heat on an intercepting body. Ultraviolet rays tend to be sent out only by objects that are at white heat and also invisible to the eye, but capable of producing hotter temperatures.

(b)

Figure 6-3 The transfer of heat by convection and radiation. (*a*) Heat transfer at an uncontrolled fire through convection. Even the small uncontrolled fire can raise room temperatures appreciably by heating the air which in turn heats the room contents. (*b*) Damage to exposure by radiant heat waves.

(*a*)   (*b*)

Radiant energy can be blocked by a reflective material that bounces back the heat energy rays. A dark surface absorbs the rays, a light surface reflects it. The person close to a fire experiences the sensation of radiant heat. There is some horizontal transference of heat through convection, but the burning sensation of heat on the flesh or the melting of asphalt shingles on an adjoining building is in the main caused by radiant heat waves.

Since heat is a product of combustion and also a necessary ingredient to the oxidation process, what is its role in the chemical reaction known as fire? Some further definitions regarding heat must be given before answering this question.

THE IGNITION PROCESS

Ignition Temperature

Ignition temperature is defined as the lowest temperature to which a substance must be heated for it to continue burning without an outside source of heat. This substance can be in either its solid, liquid, or gaseous state. With but few exceptions, oxygen or air must also be present. Initially, an exterior source of heat must be present to raise the temperature of the substance. After the ignition temperature is reached, the substance will continue to burn independently of the exterior heat source.

While the ignition temperature is more-or-less fixed for a given substance, many factors can affect actual combustion. Some of them will be discussed later in this chapter.

Testing methods of different laboratories also can bring variances in the ignition temperatures assigned by the laboratories.

We know from experience that different substances require varying intensities of heat for combustion to be initiated. (See Table 6-1.) The heat of a match will start paper burning, and the paper will normally continue to burn when the match is removed. A match may start a stick of wood burning, but the fire usually goes out when the match flame is removed. (See Table 6-1.) The wood does not reach a temperature hot enough for it to continue burning of itself. The heat of the match lights the paper—the heat of the burning paper will usually set wood afire. The heat of burning paper and wood is usually necessary to start coal burning. All have different ignition temperatures.

Other factors also enter into the condition that brings on self-sustained combustion. The mass and shape of the material play a role. Wood shavings are easier to ignite than a large log. A solid piece of wood can absorb more heat or may dissipate the heat before the ignition temperature is reached. The rate at which heat travels through various materials, the heat conductivity, also varies considerably. A piece of burning wood can be held in the hand by the end opposite the flame a long time before the temperature rises at the unlit end. On the other hand, heat travels with comparative rapidity through metals. The moisture content of cellulose materials greatly affects ignition. Low humidity brought on by a long dry spell

Table 6-1 Ignition Temperatures of Ordinary Solid Combustibles

Ignition temperatures can only be an approximation because of the variables encountered. Moisture content, mass, shape, conductivity, and duration of exposure to heat all enter into the individual case.

MATERIAL	SELF-IGNITION TEMPERATURE
Paper newsprint	446°F
Cotton batting	446°F
Cotton sheeting	404°F
Woolen blanket	401°F
Viscose rayon	536°F
Wood fiberboard	421°F–444°F
Cane fiberboard	464°F

Wood, in general, will ignite if subjected to the following temperatures for the stated periods of time.

TEMPERATURE	TIME
600°F	5 minutes
500°F	10 minutes
425°F	20 minutes
375°F	30 minutes

Source: Tests performed by the National Bureau of Standards.

constitutes a fire hazard potential. The rate and duration of application of the heat energy source enter the picture also. Wood subjected to a temperature of 500°F, a temperature below its normal ignition temperature, could ignite in ten minutes. Wood subjected to a temperature of 200°F for a long period has been known to ignite.

Records show that wood in contact with steam pipes of around 190°F for a long time will eventually break out in flame. Heat conduction through as much as 2 feet of concrete, with temperatures much less than the ignition temperature of the wood subfloor below, over a long period of time can cause ignition to occur in the wooden subfloor. Temperatures that are lower than the normal ignition temperature of wood char the wood surface and produce a chemical change that makes the substance more susceptible to ignition at lower temperatures.

Pyrolysis

With ignition, the chemical reaction has evolved through a process known as pyrolysis. As the ignition source is first applied to the combustible substance, the first chemical change takes place. Water vapor and other gases, many of them combustible, are released from the substance. The combustible gases in this stage of pyrolysis do not burn. Rather, the ignition source chars, for example, the wood surface and tends to char deeper as the ignition source remains in proximity to the surface.

As the charring continues, more gases and water vapor are released. The combustible gases ignite at a certain point in temperature. This is the ignition temperature of the substance. Burning now is all in the gaseous vapor being released from the chemical reaction process occurring in the substance from the initial temperature rise. This burning of the released vapors increases the temperature of the substance. The charred surface soon glows; a chain reaction is started in which oxygen enters into the seat of the charring area and lends support to actual combustion of the substance itself.

This somewhat simplified explanation of the pyrolysis process serves to exemplify how combustion is initiated. First comes the generation of gases as the substance breaks down; then the gases ignite. Their burning produces more heat and brings about actual combustion, or a rapid decomposition of the substance.

Ignition temperatures are important to the firefighter. At this point the pyrolysis reaction begins, and the ignition temperatures of some flammable liquids and gases can be quite low (as will be seen later). Carbon bisulfide, for example, has an ignition temperature so low that the heat of a nearby light bulb can cause the substance to ignite.

Ignition temperatures of the contents of the average home are well above the upper range of normal living temperatures, but they can be easily reached if a small, uncontrolled fire starts.

Spontaneous Ignition

Certain combustible materials are capable of attaining an increase in temperature without exposure to heat from an exterior source. This spontaneous heating can go on until the ignition temperature of the substance is reached and spontaneous ignition occurs. In some cases, the substance itself may not reach its own

Table 6-2 Some Common Substances Subject to Spontaneous Heating and the Conditions Most Apt to Bring It On

SUBSTANCE	FACTORS CONTRIBUTING TO HEATING
Charcoal	Wetting and insufficient ventilation
Foam rubber	Preheating
Cod liver oil	Oil-soaked organic materials and poor ventilation
Colors in oil	Same
Cornmeal feeds	Improper processing, excess oils
Fishmeal	Excess storage temperatures (over 100°F), improper moisture content (6 to 12 percent being ideal)
Fish oil	Oil-soaked porous or fibrous materials
Fish scrap	Loaded or stored before cooling
Linseed oil	Oil-soaked rags and fabrics, improper ventilation
Varnished fabrics	Poor ventilation, not thoroughly dry

Source: From table prepared by the National Fire Protection Association, Committee on Spontaneous Heating and Ignition.

MOIST OR GREEN HAY, GRAIN, SILAGE

ignition temperature, but by spontaneous heating, it may attain a temperature higher than the ignition temperature of an adjoining substance. The second substance will then be ignited by conduction of the spontaneous heat being generated. Some substances that are subject to spontaneous heating, and some conditions that bring it on, are listed in Table 6-2.

The process of spontaneous heating normally will not occur unless the material has become wet with water or oils or has been subjected to a low heat-energy source for a sufficient length of time. However, important factors in the process are the rate at which the heat is being generated, the air available to support combustion, and the insulation properties of the surrounding material. Oxygen of a certain proportion must be available. Tightly bundled or baled material that could spontaneously heat will not do so for lack of oxygen. Where air circulation is abundant, the generated heat will be dissipated. As an example, linseed oil has a high potential for spontaneous heating. However, if a rag soaked with linseed oil is exposed to air circulation, the heat generated will be dissipated. If the rag is thrown carelessly in a pile, the generated heat will build up to the ignition temperature. In general, oil-soaked organic materials will heat spontaneously. For safety, they must be stored in a closed container, or well-ventilated and dried before being stored.

Table 6-3 Calorific Values of Ordinary Substances

These figures are approximations derived from several sources and serve in the calculation of estimated fire loading of structures.

SUBSTANCE	AVERAGE CALORIFIC VALUE IN BTU PER POUND
Wood	8,500
Cotton	7,200
Silk	9,000
Wool	9,000
Cellulose	7,500
Paper	7,500
Charcoal	13,000
Coal	13,000–14,000
Gases (propane, butane)	21,500
Flammable liquids	16,000

Wet hay or hay not completely cured will generate heat. However, if the hay is in tight bales, enough oxygen usually will not get to the center of the bale to support heat build-up.

Some materials, such as foam rubber, that have been preheated, that is, subjected to higher than normal temperatures, may continue to heat spontaneously after the additional heat has been removed.

Calorific Content

As ignition takes place and combustion continues, heat is given off. Different substances will release differing amounts of heat dependent upon their atomic structure. This released heat is known as its heat of combustion, or calorific value. Calorific values are usually expressed in Btu per pound. Btu is the abbreviation of British thermal unit and is defined as the amount of heat necessary to raise the temperature of one pound of water 1°Fahrenheit. Under the metric system, the unit of measurement becomes a calorie, defined as the amount of heat necessary to raise the temperature of one gram of water 1° Celsius (formerly Centigrade). The calorific values of some common substances are given in Table 6-3.

In some instances the firefighter is interested in the potential fire load of a particular building or occupancy. Simply stated, the fire load is the amount of combustible materials in the potential fire area—that is, the structural components of the building and the building contents. The fire load is expressed in calorific content and can be determined by estimating the weight of the separate materials involved, and then multiplying the estimate by the approximate Btu per pound for each material. Simple addition will then supply the approximate calorific content.

To simplify fire-load calculation, an average calorific value is taken. Wood, paper, cloth, and other ordinary combustibles average 8,000 Btu per pound. Rubber, oil, and other flammable liquids are averaged at 16,000 Btu per pound. The fire load is of particular importance to the firefighter in its potential contribution to the severity of the fire to be encountered.

Flashover

It has been stated that the average combustibles in the average room have an ignition temperature of 700° to 850°F.

The standard time-temperature curve used in the testing of building components plots temperatures reached in a given time in a brick, wood-joisted building with normal contents involved in fire. Temperatures reached are 1000°F in five minutes and 1300°F in ten minutes. (See Chapter 7 for further discussion of the time-temperature curve.) Smoke, in being generated, reaches temperatures of 1200° to 1300°F. It follows that the room and its contents will soon reach their respective ignition temperatures. If there is a sufficient source of oxygen, the firefighter may witness the phenomenon of the entire room seeming to burst into fire all at once. This is known as flashover—combustible materials have reached ignition temperatures, pyrolysis starts, gas vapors are released and simultaneously flash over into fire.

Back Draft

When the room and its contents are heated up to the ignition temperature but there is an insufficiency of air or oxygen, there will be no flashover. The third side of the simplistic triangle, oxygen, is lacking and there can be no immediate fire. This poses a very dangerous situation to the firefighter. If he forces entry to get to the seat of the fire generating the smoke and heat, he will also introduce a source of oxygen. The oxygen-starved fire will reach out toward the oxygen source. The flammable vapors being released at the ignition temperature will now flash over into fire. The firefighter may be engulfed in flames or actually blown out of the burning room.

This well illustrates the need for proper ventilation, which will be discussed in Chapters 10 and 11.

STAGES OF A FIRE

Once a combustible material reaches its ignition temperature and combustion is initiated, the fire may progress, although not necessarily, through four stages.

Stage 1

This can be called the incipient stage or the beginning phase. Combustion has begun and there is an ample source of oxygen. Products of combustion are being released; water vapor, H_2O; carbon dioxide, CO_2; in some cases, sulfur dioxide, SO_2; and perhaps more. Because of the availability of oxygen, little carbon monoxide, CO, is present, although some CO is produced because of a lack of heat. Heat is building up. The temperature at the seat of the fire may have reached 1000°F, but room temperatures are still close to normal or, at the maximum, not much over 100°F.

Stage 2

Stage 2 may be classified as a free-burning stage. Room temperature will continue to rise and may have reached the vicinity of 1300°F. This will be well above the ignition temperature of the normal room content and the entire contents of the room will become totally enveloped in fire through flashover. Temperatures may reach 1550°F in a half-hour and 1700°F in one hour (the standard time-temperature curve). Smoke and gases will have increased. Because of the state of the uncontrolled burning and incomplete combustion of some materials, CO will now be present as well as the other gases discussed for normal combustible materials. Smoke and gases will have traveled to other rooms and spread vertically to the floors above. Due to the heat, gases, and smoke, entry to the fire room may be impossible. If the fire burns through to the outside and is self-ventilating, or if it is receiving sufficient air, heat build-up and fire will increase until all combustibles are consumed. However, if the room is "airtight," oxygen content will fall before all combustibles in the room are involved. The third stage will have been reached.

Stage 3

Stage 3 is the smouldering phase. Oxygen content will have dropped to near the 15 percent level. More carbon monoxide, CO, will be generated in place of carbon dioxide, CO_2. Free carbon, unburned gases, and smoke will completely fill the room. Room temperature will hold near 1000°F because of less free burning. This is still well above the ignition temperature of the average room contents. The room is completely smoke-filled. Dense smoke continues to fill the upper levels and mushrooming will be taking place. (See Figure 6-4.) If the room is airtight enough, it is possible that oxygen content will be lowered to a point where the fire is smothered out. (This does happen in rare instances.) However, in many cases the heat will crack a window which will admit ventilation, or the fire may burn through a wall or roof, thus providing ventilation. With the introduction of air or oxygen, flashover will take place, possibly a back-draft explosion, and free burning will now be evident.

Stage 4

With uncontrolled fire and sufficient oxygen present, a building and its contents will become totally involved in fire. For all practical purposes, they are a total loss even though proper fire-fighting techniques have effected extinguishment before total building collapse.

PRODUCTS OF COMBUSTION

Four products of combustion have been mentioned in the discussion of the stages of a fire. These products, smoke, gases, heat, and flame, are of great concern to the firefighter.

Figure 6-4 Smoke mushrooming. Heated smoke tends to rise to the highest level, taking a "mushroom" shape.

Smoke

In a controlled fire situation, usually sufficient air or oxygen will be introduced into the burning object or area so that very little smoke is emitted. The uncontrolled fire often has a deficiency of oxygen or air in relation to the potential fire load. With a shortage of oxygen, there will be smouldering and the emission of heavy smoke. Combustion is not complete because of the lack of sufficient amounts of the third side of the simplistic fire triangle—air or oxygen. Although there is sufficient air present for some combustion, some of the fuel is vented into the atmosphere in the form of tars, resins, and carbon products, commonly called smoke. The combustible material has been subjected to heat. It has partially decomposed and has broken down into elements or other compounds. That emission known as smoke has absorbed some of the heat. The temperature of this smoke can reach between 1000°F and 1200°F. The heating effect causes the smoke to rise until the upper ceiling or roof line prevents further travel. The smoke will then tend to mushroom and to spread out horizontally, filling the upper floors.

Smoke does not hold heat very well, but tends to give it off through conduction to the walls, ceilings, and contents of a structure. If this conduction continues, they will be heated sufficiently to reach ignition temperatures.

The firefighter soon learns to identify the odors of smoke given off by varying materials. The special odor from a deep-seated wood fire will indicate that a "good" fire is ahead of him even before he has observed the flame or felt the heat. Characteristic odors can reveal whether a mattress or upholstered furniture, a pan of meat on the stove, or an electrical appliance is burning.

These unburned particles of matter—smoke—may be an irritant to the firefighter. Exposure causes eye irritation plus irritation to the mucous membranes of the respiratory tract of nostrils and throat. Many toxic gases also exist in the composition of smoke and contribute further hazards to the firefighter. Smoke is responsible for nine out of ten injuries to the firefighter, and is classified under the broad term *smoke inhalation.*

Toxic Gases

Certain toxic gases may be emitted in the process of combustion.

CARBON MONOXIDE—CO

Carbon monoxide is a colorless, odorless, tasteless gas. The flammable limit range is 12 to 75 percent.[1] CO is slightly lighter than air, the vapor density being 0.967. Carbon monoxide is always a product of incomplete combustion of organic materials, those containing carbon, oxygen, and hydrogen. When large volumes of smoke are present and organic materials are burning, carbon monoxide almost certainly is present also. It has an ignition temperature of 1128°F. Recall that the temperature of smoke can reach 1000°F to 1200°F. It is always possible that carbon monoxide, with an explosive range of 12 to 75 percent, may itself contribute to the fire spread. Furthermore, concentrations much less than this are harmful to the firefighter. Lesser concentrations may cause a headache at the minimum; with long exposure, death is possible.

CARBON DIOXIDE—CO_2

Carbon dioxide is a colorless, odorless gas that is nonflammable and nonpoisonous. It is heavier than air, its vapor density being 1.529. CO_2 will not support combustion or life. It is a product of the combustion of fuel and is also given off in the exhalation of air in the breathing process. If introduced into the air in any appreciable concentration, it lowers the percentage of the oxygen content and may cause possible suffocation due to lack of oxygen. (It is this smothering effect that is used to advantage in the use of CO_2 as a fire-extinguishing agent.) Being heavier than air, it settles into the lower level of a fire building, into manholes, wells, sewers, or mines. A concentration of 0.1 to 0.5 percent can produce symptoms of weakness and headache. Concentrations of 8 to 9 percent may cause suffocation through asphyxia, although death from CO_2 is uncommon.

The inherent hazard in CO_2 is that it overstimulates the breathing system. Small concentrations of CO_2 have a stimulating effect on the respiratory system and thus cause more rapid and deeper breathing. A concentration of 2 percent increases the normal breathing rate 50 percent; 3 percent will increase the breathing rate 100 percent. This deeper, faster breathing will also contribute to the absorption of other toxic gases into the system more rapidly.

[1]Chapter 12 contains a detailed explanation of flammable limit and flashpoint.

SULFUR DIOXIDE—SO_2

SO_2 is a nonflammable, colorless, tasteless, poisonous, and highly irritating gas. It is much heavier than air, with a vapor density of 2.264. In addition to being extremely toxic, its strong sulfurous odor is so irritating and noxious that a person cannot breathe in it and will seek escape. If escape is delayed, death may result quickly. Slight exposure can cause later symptoms of respiratory damage, including bronchopneumonia.

HYDROGEN SULFIDE—H_2S

Hydrogen sulfide, H_2S, is a gas more poisonous than carbon monoxide. It is slightly heavier than air, having a vapor density of 1.2. In lower concentrations, it is identified by its characteristic "rotten egg" smell, but in higher concentrations this sense of smell identification is somewhat lost and the victim is unaware of the presence of the gas. Damage to the central nervous system may bring on paralysis and possible death. A concentration of 0.04 to 0.07 percent for a half-hour can cause symptoms of dizziness, dryness, and intestinal disturbance, as well as pain in the respiratory system. Greater concentrations can result in paralysis.

AMMONIA—NH_3

Ammonia, NH_3, is a flammable, colorless, poisonous gas with a sharp penetrating odor and taste. Lighter than air, it has a vapor density of 0.597; the flammable limit range is 15 to 26 percent in air. Exposure to 0.25 to 0.65 percent concentration for one half-hour can be fatal.

However, because of the pungent, irritating effect on the eyes, nose, throat, and lungs, the victim will flee the atmosphere containing ammonia before serious effects can result. Ammonia has a great affinity for water. A concentration of ammonia vapors in the atmosphere can be absorbed by the use of water spray.

HYDROGEN CYANIDE—HCN

Hydrogen cyanide, HCN, is a highly toxic gas, flammable, tasteless, very poisonous, and with an odor of bitter almonds. It is lighter than air, with a vapor density of 0.697. The flammable limit ranges from 5 to 40 percent in air. Exposure to 0.3 percent is fatal (HCN is used in gas-chamber execution). Fortunately, it is not usually produced in dangerous quantities in the average fire.

ACROLEIN—C₃H₄O

Acrolein, or acryllic aldehyde, is a toxic gas given off during the combustion of petroleum products, fat, oils, grease, and many other common combustible materials. It is heavier than air, with vapor density of 1.9. The flammable limits are 2.8 to 31 percent in air. It is usually given off in low concentrations, but can be deadly in concentrations of 10 parts per million.

GASES RELEASED BY SPECIFIC MATERIALS

The foregoing gases are some of the more common hazards to be met in the process of combustion of various materials. The following gases are characteristic of specific substances.

Wood, cotton, and paper products all contain cellulose and are the chief fuel in most fires. In uncontrolled combustion there quite often is combustion without sufficient oxygen. Here, the byproducts of combustion may be carbon monoxide, formaldehyde, formic acid, carbolic acid, methyl alcohol, acetic acid, and other compounds. Smoke will be thick and heavy with much particulate matter and irritants. When the oxygen supply is increased, combustion will be more complete. As a result, only small amounts of organic compounds, along with carbon monoxide and carbon dioxide, will be found as byproducts of combustion.

Plastics give off, as byproducts, carbon monoxide, hydrochloric acid, cyanide, nitrogen oxides, and other toxic compounds. Much depends on the type of resin used in the plastic. In a general sense, the hazard is no greater than that of other common materials.

Burning rubber gives off carbon monoxide, hydrogen sulfide, and sulfur dioxide, all with a heavy black smoke; a further byproduct may be the type of sick headache that may last several days.

Burning silk may be responsible for emitting the deadly hydrogen cyanide gas and the toxic ammonia.

Burning wool gives off as byproducts carbon monoxide, hydrogen sulfide, sulfur dioxide, and hydrogen cyanide.

Petroleum products generate a dense black smoke. Acrolein, carbon monoxide, and carbon dioxide are present in this heavy smoke.

This list is not all-inclusive, but is given in the main only to indicate the mixture of gases possible in the process of combustion. The toxicity of the combination of several of these gases in combustion byproducts is greater than the toxicity of any single gas.

Most gases are highly irritating and nauseating. The exposed person will flee the environment if possible—usually before lasting harm can be done.

Specific industrial and chemical hazards will be discussed in Chapter 12.

Heat

The firefighter exposed to high temperatures may suffer heat exhaustion, dehydration, burns, thermal injury to the respiratory tract, and an increased heartbeat. Consequences range from minor irritation to death. The effect of moist heat is more deleterious than hot, dry air.

The firefighter operating at the fire scene must wear fully protective clothing. The effects of heavy turnout clothes, plus the heat of the fire on a hot summer day, may very well cause the firefighter to suffer from dehydration. He must take care to maintain healthful levels of salt and water in the body and to exercise caution in extreme heat conditions.

A temperature of 300°F is said to be the highest temperature that man can breathe, and this only for a very brief time and only in an atmosphere of dry air. Moist air at this temperature can prove fatal even if exposure is for only a short period.

However, a person's body, when exposed to heated air, will experience pain and discomfort in sufficient degree to cause the person to flee the environment before his or her entire body collapses.

Flame

Flame will normally be present in freely burning materials. It is the flame that provides the luminous lighting quality that accompanies fire. When flame is seen, it is proof of the presence of fire and indicative of its intensity.

Flame is also responsible for most of the burns suffered, although burns without flame can occur from the heat of combustion and the conduction of heat.

The human body is covered with what is known as skin. Skin itself is made of two layers, the dermis and, above it, the epidermis. The epidermis is a very thin outer layer less than one-tenth the thickness of ordinary tissue paper. Yet it is this thin layer that helps the body retain moisture and protects the body from bacterial infection and the loss of fluid.

Burns are now classified into four categories: first-, second-, third-, and fourth-degree burns.

First-degree burns affect only the outer layer of the epidermis and do not penetrate this thin layer, although they may cause redness, pain, and perhaps an accumulation of fluid at the burn area.

Second-degree burns penetrate into the epidermis and form blisters with subcutaneous fluid accumulating.

Third-degree burns penetrate into the dermis and leave a dry-charred area where the burn occurred. Nerve endings may also have been affected, but with resulting relief from the feeling of pain. In most instances, the victim who suffers third-degree burns over a small body area, or a deep-seated burn caused by burning clothing or long contact with flame, is more liable to die than the flash-fire victim with burns over a greater part of his or her body.

Fourth-degree burns are third-degree burns that appear to be only a first-degree type. This is a flash burn with a deep penetration not at first suspected.

The flash-fire burn, while very painful and disfiguring, is the result of a momentary flash and usually results in only a first- or second-degree burn which does not penetrate into the dermis layer. When this type of burn involves deep penetration, it is as serious as the more obvious third-degree burn. Burns over 65 percent of the body are usually fatal. Scarred tissue and skin can be replaced by skin grafting, but death is most often due to infection and loss of body fluid.

This overview indicates briefly the health hazards faced by the firefighter. Smoke conditions, heat, lack of oxygen, and exposure to toxic gases are all factors in the classification of smoke inhalation to which the firefighter is vulnerable. Statistics seem to indicate that three out of ten injured firefighters suffer from injuries coming under this broad classification.

In addition to smoke inhalation, the firefighter is subjected to accidents that result in cuts, sprains, broken bones, or foreign objects in the eye. The physical labor under the stress of excitement and anxiety may lead to heart disease and exhaustion.

Figure 6-5   Firefighter injuries and their severity.

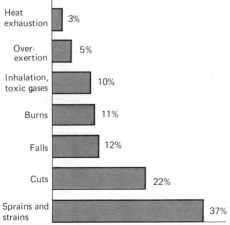

Courtesy of the International Firefighters Union.

Statistics for 1970 reveal that 38,583 firefighters were injured, over 25 percent of them also lost time from work. (See Figure 6-5.)

The firefighter also has the largest death rate of all occupations. In the United States in 1970, 115 firefighters died in the line of duty per 100,000 firefighters.

SUMMARY

The process of combustion follows the basic laws of the natural sciences. The scientific terms and definitions must be understood by the firefighter if he is to understand fire.

Heat is a form of energy. It is a measurement of the molecular motion in a substance. Heat energy comes from several sources:

1. Chemical oxidation is the heat released in chemical reactions, and when rapid enough, it causes combustion.

2. Electrical heat energy is caused by arcing, dielectric heating, induction heating, or resistance to current flow. Static electricity and lightning give an arcing action between two bodies of opposite charge.

3. Mechanical heat energy is caused by the heat of friction and resistance to motion of two bodies in contact.

4. Nuclear heat energy is the release of energy from the nucleus of the atom.

Heat reaction can be exothermic or endothermic. In an exothermic reaction, heat is given up by the substance. In an endothermic reaction, heat enters the substance.

Heat is transferred from one substance to another in one of three ways.

Conduction is the passage of heat that balances out the existing temperature differential of substances touching one another.

Convection is the passage of heat from one body to another through a medium of transfer, such as a liquid or a gas.

Radiation is the transfer of heat energy by infrared rays, visible light, and ultraviolet rays.

Ignition temperature is the lowest temperature to which a substance must be heated for that substance to continue burning without an outside source of heat. Ignition temperatures vary from one substance to another and in the same substance dependent on moisture content, mass, and shape.

As ignition temperature nears, flammable vapors are released. The surface of the substance chars. Continued heat application causes the vapors to ignite, and the fire helps increase surface temperature on the substance. The charred surface glows and also ignites. This process is known as pyrolysis.

Some substances increase in temperature without an outside source of heat. This phenomenon is known as spontaneous heating, which can lead to spontaneous ignition if the temperature increases enough.

The heat energy released from a substance can be measured in Btu per pound; this measurement gives the calorific content of the substance.

A substance that has reached its ignition temperature but has insufficient oxygen to burn will flash over into fire when oxygen is introduced. A fire explosion may result as the air-starved fire receives the oxygen.

Fire passes through several stages.

Stage 1 is the incipient or beginning phase.

Stage 2 occurs if sufficient oxygen is present so that the fire can progress to a free-burning state.

Stage 3 is the smouldering fire. In some cases oxygen is in short supply or temperatures are slow to build up and the fire will smoulder before entering the burning stage.

In stage 4, temperatures now are quite high, oxygen is plentiful, and combustion of all combustibles will be complete.

In the process of combustion, smoke, gases, heat, and flame are evolved. These are harmful and are responsible for three out of ten injuries to the firefighter.

DISCUSSION TOPICS

1. What is heat energy and what are some of its sources?

2. Discuss the methods of heat transfer.

3. Discuss ignition temperatures and factors involved in self-sustained combustion.

4. Relate flashover and back draft and their effects on the spread of fire.

5. The four products of combustion are smoke, heat, flame, and gases. Discuss these hazards and how they affect fire-fighting techniques and the firefighter.

RESEARCH PROJECTS

1. List the sources of heat energy and cite examples of these heat sources in industry.

2. List the methods of heat transference. Cite some of their uses in society and industry. Give examples of each factor in, or a cause of, an uncontrolled fire.

3. Cite examples of endothermic and exothermic reactions.

4. Using a small structure and its content for the problem, determine the calorific content in Btu.

5. The fire service as well as allied interests is concerned with the very high proportion of injuries caused by smoke inhalation. Research some measures being taken, or which should be taken, to reduce these incidences.

FURTHER READINGS

Campbell, Richard D., *College Chemistry, A Survey,* Harcourt Brace Jovanovich, New York, 1968.

Garard, Ira D., *Invitation to Chemistry,* Doubleday, Garden City, N.Y., 1969.

Patton, A. R., *Science for the Non-Scientist,* Burgess, Minneapolis, Minn., 1962.

Pauling, Linus, *College Chemistry,* 3d ed., Freeman, San Francisco, 1964.

Porter, George, *Chemistry for the Modern World,* Barnes & Noble, New York, 1962.

Vaczek, Louis, *The Enjoyment of Chemistry,* Viking, New York, 1964.

Our primitive ancestors began as hunters and food gatherers. They were obliged to range over a wide area and in small groups to supply their needs. It was only after human beings learned to control fire that they could begin a process of settlement and domestication as tillers of the soil. Thus the agricultural village was born.

Domestication brought further social and economic needs and family providers soon specialized to supply those needs—the hunter, the miner, the shepherd, the farmer, the fisherman, the woodsman, all contributed to the life style. The village expanded. By 2500 to 2000 B.C., the essential features of the urban city had been developed.

With the city came the need for codes to regulate conduct and safety. Among these codes were laws regulating building construction. Fire prevention ordinances followed.

BUILDING CODES AND FIRE ORDINANCES

At the present time, building codes and fire prevention ordinances exist on the national, state, and local levels.

In brief, a building code is concerned with the safety of life and property through regulation and control of building design, construction, materials of construction, location, occupancy, and certain equipment used. A fire prevention code, also concerned with life, property, and the public welfare, regulates the storage, use, and handling of hazardous materials and the processes and equipment used. The code also deals with the maintenance of buildings and premises and those factors affecting life safety.

The power to establish a building and fire prevention code is usually granted the municipality under its charter from the state.

BUILDING
CONSTRUCTION

7

Local practices, customs, and climatic conditions bring variations in building codes throughout the country. Guidelines to the establishment of an effective code have been offered through five model building codes. These model codes and the organizations offering them are:

National Building Code, of the American Insurance Association

Uniform Building Code, of the International Conference of Building Officials

Southern Building Code, of the Southern Building Code Congress

Basic Building Code, of the Building Officials Conference of America

National Building Code of Canada, of the National Research Council of Canada

In addition to these model codes, there is a long list of technical standards published by various interests, many previously mentioned, that can be incorporated into a building code. Many of these standards, or portions of the model codes, are adopted by "reference." That is, the local building code states that the particular standard is on file and is to be considered a part of the local code.

The modern building code is a performance code. Standards of performance are established and any material that meets these standards can be used as a component of the building structure.

Earlier codes had a tendency to specify materials rather than performance. This negated the use of newer materials that could perform as well as, or better than, the older building materials.

In the larger municipalities, the fire prevention code is administered by the fire department. The larger departments handle fire prevention administration as a separate bureau. (See the organization chart in Chapter 4, Table 4-2.)

Both the building department and the fire prevention bureau need qualified, experienced people to properly regulate and enforce all the technical aspects of life safety that can arise, given the economy and technology of today.

All building codes recognize several types of building construction. These have varying degrees of fire resistance. Most building codes, although there are some variations, classify five standard types of building construction. (The code used in Western states does not classify in this manner.)

1. Fire-resistive

2. Noncombustible

3. Heavy timber

4. Ordinary

5. Wood frame

These five standard types of building construction are defined in National Fire Protection Association Standard No. 220. Minimum performance standards are also established.

Older codes carried the classification of "fireproof building," but fires in many buildings ranked as fireproof gave vivid evidence that no structure is truly untouchable by fire. A fireproof building will contain combustible contents and some combustible components, such as wood trim, doors, and perhaps paneling. Even those structural members previously listed as fireproof were subject to some reaction if exposed to sufficient heat. For these reasons, the "fireproof" classification has been amended to "fire-resistive."

FIRE TESTING OF BUILDING CONSTRUCTION AND MATERIALS

In discussing types of building construction and building materials, a fire-resistance rating in terms of hours of resistance has been given. These ratings can be one hour, two hours, or many hours of resistance before the materials succumb to the effects of the fire. A natural question then arises: What are the criteria used to determine this rating?

In 1918 a conference of representatives from eleven technical sources met to formulate a rate-of-temperature-rise curve that would be indicative of the temperature reached at the normal fire. Many factors had to be considered, and of course any standard derived would tend to be an "average," since many fires show a much smaller temperature rise and some fires exhibit a faster rise and higher temperatures.

Professor Ira Woolson, consulting engineer of the National Bureau of Fire Underwriters (NBFU), had a strong influence on the determination of the characteristics of the standard time-temperature curve. This curve (Figure 7-1) is now used as the standard for testing building materials and components.

Some points on the time-temperature curve were judged to characterize the average fire and were established as the standard to be used in the testing of materials.

Heat of 1000°F at five minutes, which is above the average ignition temperature of the average room content.

Heat of 1300°F at ten minutes, which, it can be seen, is a very rapid temperature rise in the first ten minutes. At the end of ten minutes,

Figure 7-1 Standard time-temperature curve.

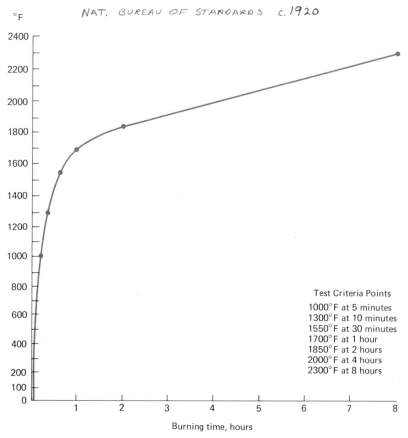

Test Criteria Points

1000°F at 5 minutes
1300°F at 10 minutes
1550°F at 30 minutes
1700°F at 1 hour
1850°F at 2 hours
2000°F at 4 hours
2300°F at 8 hours

Burning time, hours

average combustibles have been heated up enough for flashover to occur if sufficient oxygen is present. This fact indicates the importance of time to the firefighter and the need to arrive at the fire scene before the building and its contents can reach these temperatures. The first few minutes are critical! The fire may outspeed the response.

Heat of 1550°F at thirty minutes, showing that the temperature rise has now leveled off somewhat although it will continue to rise. Material that has not ignited at this time and at this temperature is said to have, at a minimum, a half-hour fire-resistance rating.

The NFPA, the American Society for Testing and Materials (ASTM), and the Underwriters' Laboratories, Inc., have published similar texts detailing the standard methods of fire testing of building construction and materials. The materials or components to be tested are subjected to controlled temperatures that follow the time-temperature curve for a duration of time equal to the specific period that the material or components are to be rated, or until the materials ignite.

The test specimen, under the time-temperature conditions necessary for acceptance, must bear working loads and stresses to which the material will be subjected in field use. In addition, walls, floors, and roof must be able to retard the flow of heat through them so that cotton waste on the exterior side cannot be ignited by flame or gases, or so that the temperature rise on the unexposed side will not be more than 250°F.

An additional required test may be the application of a hose stream on a test specimen that has been subjected to the time-temperature factor of one-half the rating period. Because of rapid cooling, there is possible structural failure of some materials that are struck with a hose stream even though they apparently can withstand the fire-endurance restrictions.

Observations on the completed tests will include structural behavior or change in the tested material, such as deformation, cracking, spalling (chipping or scaling), and smoke production.

TYPES OF CONSTRUCTION

With the understanding of how materials are tested and classified a discussion of the five standard building types, as defined in NFPA Standard No. 220, is worthwhile.

Table 7-1   Fire-Resistance Rating of Structural Members in Hours as Recommended in Fire Administration Reports Standard No. 220

STRUCTURAL MEMBERS	CLASSIFICATION	
	3-hour	2-hour
Bearing walls or bearing portions of walls, exterior or interior (all bearing walls and bearing partitions must have adequate stability under fire conditions in addition to the specified fire-resistance rating)	4	3
Nonbearing walls or portions of walls, exterior or interior, must be noncombustible (NC). Fire resistance may be required in such walls by conditions such as fire exposure, location with respect to lot lines, occupancy, or other pertinent conditions	NC	NC
Principal supporting members, including columns, trusses, girders, and beams for one floor or roof only	3	2
Principal supporting members including columns, trusses, girders, and beams for more than one floor or roof	4	3
Secondary floor construction members, such as beams, slabs, and joists not affecting the building's stability	3	2
Secondary roof construction members, such as beams, purlins, and slabs not affecting the building's stability	2	1½
Interior partitions enclosing stairways and other openings through floors	2	2
(One-hour noncombustible partitions may be permitted under certain conditions)		

Fire-Resistive Construction

Fire-resistive construction requires that all structural members, such as walls, partitions, columns, and floor and roof construction, are of noncombustible materials with fire-resistance ratings not less than those specified in Table 7-1.

A clear understanding of these requirements is not possible without a definition of the term "noncombustible." The word has been applied to materials that will not ignite and burn when subjected to fire. It is well again to point out that this fire-resistive quality does not negate the possibility of structural failure if the material is subjected to fire. Steel, brickwork, and concrete, all noncombustible materials, provide the skeletal structural strength of the fire-resistant building. Yet they can fail structurally if exposed to sufficiently high temperatures for a long enough period of time. These same materials will fail even faster if struck by a hose stream after being subjected to the heat of a normal fire.

Curtain walls of brick and windows often complete the building facade. This type of construction has permitted building heights to rise higher and higher. Before his death in 1959, the noted architect Frank Lloyd Wright spoke of building the mile-high structure around a central supporting core of reinforced concrete.

A current trend is to the all-glass facade or a solid curtain wall unbroken by any opening. This new construction, although the most fire-resistant to date, raises problems for the fire service— problems that have yet to be completely resolved. There is the difficulty of getting to the fire floor— elevators are frequently tied up on other floors and emergency elevators are a rarity. Ventilation of the fire floor is a tricky and difficult task. Water supply must depend upon standpipes and adequate pressure to deliver the water to the fire floor. Evacuation of the building's occupants is a slow and difficult task. Fire protection engineering recommendations are often bypassed. There is a time lapse between the introduction of new building features and new building code ordinances.

The fire-resistive building need not be a skyscraper or a high-rise, however. There are many fire-resistive buildings that are only one or a few stories high, and many have low cubic-foot content. In these structures, it is the fire loading and occupancy that determine the possible fire severity.

Noncombustible Construction

Noncombustible construction is that construction in which all structural elements are of noncombustible materials, but which are not generally fire-resistive. This means that, although the structural members will not burn, there may be a structural failure due to heat. This type of construction is identified by the exposed steel of the framing members. Structural columns and beams normally will be of structural steel. Enclosing exterior walls may be of curtain-wall construction. If columns and beams are covered for fire protection,

Figure 7-2   Collapse of unprotected steel girders exposed to fire. Unprotected steel starts to fail at 1000°F to 1400°F, a temperature reached in five to fifteen minutes on the standard time-temperature curve.

Courtesy of the Chicago Fire Department.

they will be only those of the exterior walls. Ceiling and roof construction may be of steel open-web truss construction with roof covering of poured gypsum.

The inherent danger in the exposed steel noncombustible structure is the danger of structural collapse. When subjected to heat, steel beams can expand sufficiently to push out exterior walls. The steel beam or column that is under extreme heat of a fire has been known to buckle and collapse when struck with a hose stream or when the period of fire exposure is long enough. Figure 7-2 illustrates one such collapse.

While the building structure itself may be noncombustible, some components, such as doors, trim, or a grid type of ceiling with acoustical panels, may be combustible. Again, the fire-loading factor of the building contents may offer a high hazard potential. This type of construction may be employed in the shopping center with high fire-load contents. An example at the other extreme is the metal-fabricating or metal-storage warehouse with only a small occupancy fire potential.

Heavy Timber Construction

The bearing walls of heavy timber construction must be of noncombustible material with a minimum fire resistance of two hours and must remain stable under fire conditions. Nonbearing exterior walls must be of noncombustible materials. Occupancy, fire exposure, or other factors that can affect fire conditions can cause an increase in the two-hour fire-resistance rating of exterior bearing and nonbearing walls.

Columns, beams, and girders may be of wood. If they are, the columns must be of at least 8-inch timbers, and the beams and girders must be of at least 6- by 8-inch timbers. If not of wood, these members must be of a fire-resistant material carrying a one-hour fire-resistant rating.

Subflooring normally is of tongue-and-groove wood planking with a minimum 3-inch thickness. The finish floor is of 1-inch flooring laid at right angles or diagonal to the subfloor.

Beams and girders of roof-framing members supporting roof loads must be at least 6 inches in their smallest dimensions. Timber truss or arches may be used for roof framing. Trusses or arches can be built up of smaller members, not less than 3 inches of wood blocked solidly, but the overall dimensions of trusses must be a minimum of 4 inches by 6 inches. Roof decking must be not less than a minimum of 2 inches in thickness.

Interior partitions around openings and stairways must have at least a one-hour fire-resistance rating.

Heavy timber or "mill" construction results in a "slow-burning" building. It is believed superior to unprotected noncombustible steel. The wood timbers can burn, but because of the heavy dimension, they must be exposed to fire for a long period before structural collapse takes place. The experienced firefighter can gauge the amount of char and burn and predict collapse before it occurs. Neither he nor anyone else can forecast whether structural collapse is imminent by an observation of the steel beam. Collapse can be instantaneous. The heavy timber mill-constructed building is exemplified by the loft type of building or the factory building of the early twentieth century. An inherent hazard is the open wooden stairway from floor to floor.

After World War II, a renaissance in building construction took place. Heavy wooden timbers of the size and length needed for heavy timber construction were difficult to obtain. Fire-resistant construction of reinforced concrete became prevalent for many types of structures and occupancies other than the high-rise. The era of the mill-constructed building, as known, was over. There was, however, a revival of heavy timber construction in the exposed glue-laminated wooden-beam structure. Roof decking is of 2-inch planking and is left exposed. This "new look" is found in the typical A-frame structure, in churches, and auditoriums, and in many other occupancies.

Ordinary Construction

Ordinary construction requires the same type of exterior wall construction, both for bearing and nonbearing walls, as heavy timber construction. Roofs, floors, and interior partition-framing is of wood, but of smaller dimension than the interior beams and timbers of the timber construction. Interior structural members may be exposed, but more typically, the construction of wood stud and joist is covered with lath and plaster, or dry-wall construction of panel board or sheetrock.

The heavy timber construction has no hidden stud or joist space. Beams are quite widely spaced and the bottom of the flooring serves as the ceiling of the floor below. The ordinary constructed building normally has wall-stud spacing of 16-inch centering and ceiling-joist spacing of 16-inch centering. When closed in, these spaces offer a flue way for fire travel through openings that have been cut for electrical conduits, plumbing stacks or pipes, or heating ducts. It is this type of construction that must be "opened up" to detect hidden fire.

Some commercial one-story buildings of ordinary construction may have a truss roof. This roof is particularly vulnerable to fire and structural collapse. Roof boards are nominally 1-inch sheathing or ¾-inch plywood. Trusses are built up of small framing members; structural strength depends on the truss action. The smaller component parts of the truss are quickly consumed when exposed to fire and the entire truss then collapses, bringing portions of the roof down with it. Such a collapse is illustrated in Figure 7-3.

Figure 7-3   Truss roof construction. Note the small framing members of the truss still intact. It is these small component members that contributed to the collapse of the fallen trusses.

Wood-Frame Construction

When exterior walls, bearing walls, partitions, and floor and roof construction are of wood or other combustible material, the type of construction is known as wood-frame construction. This typical construction has exterior studding walls of 2- by 4-inch, or 2- by 6-inch, wood. The exterior normally is covered by ¾-inch sheathing boards. Modern construction uses ½-inch gypsum or ½-inch plywood for the sheathing coat. A finish of wood siding, plywood, aluminum, or plastic siding is the finish coat. Stucco, plaster, or a 4-inch brick veneer may also be used as the exterior coating.

Many older frame homes in the urban areas have been covered by a simulated brick or stone finish. This material is made of a wood fiberboard with a stone finish, simulating brick or stone, cemented on with asphalt binder. The radiant heat of an exposure fire causes the asphalt to liquefy and drip.

Insulation that is in itself fire-resistant, such as the spun-rock or glass type of insulation, and that has been placed between studding or ceiling joists of the wood-frame or ordinarily constructed building, tends to retard the passage of fire.

Framing of stud members in the wood-frame or ordinary construction usually follows one of two methods: balloon framing or western (platform) framing (see Figure 7-4). Balloon framing is no longer so prevalent as it once was, but it still exists in the split-level type of house. In balloon framing, the 2- by 4-inch studding runs continuously from foundation-plate line to the top-floor ceiling. Where the various levels occur, a ledger board of 1 inch by 4 or 6 inches is notched into the studding to carry the floor joists. If the studding is not properly fire-stopped at the floor levels, this type of construction offers continuous paths for fire travel to progress up partition walls or to cross over into ceiling joists and spread horizontally across the fire building. It is frequently this type of construction in the older, uninsulated home that evokes the firefighter's remark, "No matter where we opened up, there was fire."

In the western type of framing, the studding runs only from grade line, or floor line, to the ceiling line. The wall studding has a single bottom plate and a double top plate that serves as a fire stop—unless it has been cut for the pipe or duct runs that service the building.

All standard and acceptable present-day methods of building construction fall into one of the five mentioned building construc-

Figure 7-4 Details of typical wood framing. (a) Balloon framing. (b) Western, or platform, framing. Note paths for fire travel as shown by the arrow in (a), and the natural fire blocks, also marked by an arrow, in (b).

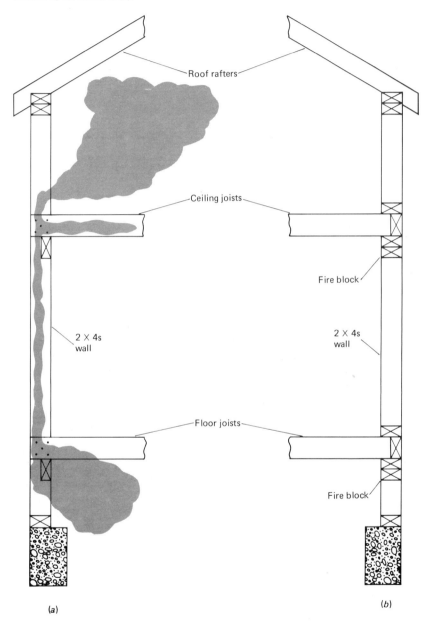

Roof rafters

Ceiling joists

Fire block

2 X 4s
wall

2 X 4s
wall

Floor joists

Fire block

(a)

(b)

tion types. There may be an interchange or a variation in form, but the finished structure can be classified into one of the five basic types. In essence, minimum standards quoted are performance-based criteria. Any material that can meet the required fire-resistance rating, that can perform the job required of it, may be used in the construction.

It then becomes a matter of aesthetics or material cost that determines the choice of material used. Thus the use of new materials, developed within the private sector, is not discriminated against once they are in compliance with the accepted standard fire tests. Only the final test of time remains to be passed before the material takes its place in building construction history.

Temporary enclosures such as tents or air-supported structures are not classified as buildings under the guidelines of NFPA No. 220. However, they do create fire protection problems involving both life and property. There is the ever-present hazard of collapse and entrapment of victims. This danger is compounded if fire is present. Many of these enclosures are used for public assembly, and their contents may be highly combustible. One example is the circus fire in Hartford, Connecticut, in 1944 that took 168 lives.

THE SPREAD OF FIRE

The normal path of fire communication is in an upward vertical direction. Heat and smoke have a tendency to rise as temperatures build up. Fire can burn its way upward much faster than it will burn through a floor and drop downward to a room below.

At the same time that it is traveling upward, especially if upward travel is blocked, it will travel in any direction horizontally. Wind velocity and direction, along with the construction and occupancy of the fire building, will determine the fire direction.

Some of the contributors to the vertical spread of fire are

Stairwells and escalators

Elevator shafts

Pipe shafts

Conveyor openings

Partitions

The goal of the fire protection engineer is to seal off each floor of a building, making it a separated unit so that the passage of heat and smoke to upper and lower floors is stopped.

For this to be done effectively, all the vertical openings to the other floors must be enclosed with a fire-resistant material commensurate with the fire-resistant rating of the interior construction. Openings in partitions enclosing stairwells and elevator shafts must be provided with fire doors. A fire door is a door designed and tested to withstand a required fire-hour resistance rating. Pipe shafts and direct openings must be sealed off at floor and ceiling. Conveyor openings can be protected by a water-spray curtain or sprinkler head that is activated under fire conditions when protection in another manner is not feasible. Escalator openings, if used as a required exit, should be enclosed as a stairwell. If an escalator exists in addition to enclosed stairwells, it may be protected by a water-spray curtain.

Fire must be prevented from traveling horizontally over an uncontrollably large space by subdividing the large horizontal area with fire walls or partitions. A fire partition is a wall subdividing only one or several floors; it does not extend from basement through roof. A fire wall subdivides areas enclosed by the exterior walls, but, in addition, it extends from basement-floor level through the roof and above. (That portion extending above is known as a parapet, illustrated in Figure 7-5.) Thus, flames breaking through the roof in one area cannot communicate to another portion of the roof without "jumping" the fire wall parapet. Exterior walls also extend above the roof line as a barrier to communication of fire by exposure.

Openings in these fire walls or partitions to provide access to other areas within the exterior enclosing walls must be protected by fire doors. Large areas must be subdivided into smaller areas so that they can become more readily controllable in the event of fire. Subdivisions will depend somewhat on construction and occupancy. There are many case histories of fires that owe their large loss to the undivided large area. The General Motors fire in Livonia, Michigan, encompassed a building spreading over three acres, all under one roof with no fire walls or partitions.[1]

Although properly subdivided with fire walls and partitions, unsuspected openings may present a hazard. Openings may have

[1]This fire occurred on August 12, 1953, in an auto transmission factory. Six persons were killed, and the monetary loss was $32 million.

been made for the passage of pipes or ducts; remodeling may have necessitated the breaching of a wall. In many instances, a common wall between two mercantile or industrial occupancies may be breached to provide a doorway so that the two occupancies can operate as one. If this opening is left unprotected, a natural path of fire communication in either direction is provided. An electrician or a plumber may have knocked an 8-inch or larger hole in a wall for a 1-inch pipe. The hole permits fire to communicate through the wall also. In setting joists on the fire wall, the carpenter may have permitted the joists of one section to butt up against the joists of the next. With no fire break between joists, fire can travel across when the joist burns.

In other cases, partitions on the floors may provide horizontal protection to fire spread, but the concealed space between ceiling and roof may have no subdivisions. In the ordinary or frame type of construction, this area will contain much wood framing — particularly in the truss-roof type of construction. Fire can spread quite rapidly in this space before it is even detected, and when detected, it is difficult to head off and control. Many times the building is "lost" before fire in the space is detected. The NFPA Life Safety Code [2] recommends a maximum of 3,000 square feet without a fire-stop partition in attic areas.

It must be realized that the occupancy of the building will play a decisive factor in the spread of fire within that given structure. A good building code is cognizant of this and will not permit more hazardous occupancies in a structure of low-hour fire-resistance rating. Neither will it allow life to be endangered by permitting large groups of people to be housed or gathered in the low-hour fire-resistance-rated structure.

Occupancies have been classified and a correlation between construction and occupancy needs can be established.

CLASSIFICATION BY OCCUPANCY

NFPA Standard No. 101, Life Safety Code, 1967 edition, also classifies occupancies and hazards of contents. The following guidelines are in the main from this code.

1. *Assembly:* Crowd attendance by persons voluntarily and not usually of a fixed attendance. Usually able-bodied persons not subject to discipline and control, with inherent panic hazard in an emergency. Fifty or more persons gathered for worship, entertainment, amusement, or awaiting transportation, for example, in churches, theaters, bowling alleys, passenger stations.
2. *Educational:* Group attendance of a fixed and regulated nature subject to discipline and control, as in schools, colleges, and kindergartens.
3. *Institutional:* Eating and sleeping facilities; care of the sick, young, aged, and physically or mentally handicapped; penal and corrective incarceration. Specifically,

Health care facilities, including hospitals and nursing homes

Residential facilities including custodial care, nurseries, and homes for aged or mentally handicapped

Residential institutions, providing restrained care in jails, reformatories

[2]NFPA Standard No. 101 contains recommended standards for safety to life from fire in buildings and structures. The document is concerned mainly with those features that will ensure safe evacuation in event of fire or other life hazard conditions rather than with what happens to the building.

4. *Residential:* All normal residential purposes with sleeping accommodations other than institutional. Subgroups include:

Hotels and motels (transient)

Apartments

Dormitories (orphanages)

Lodging or rooming houses

One- and two-family dwellings

5. *Mercantile:* Stores, markets, and other locations for the display and sale of merchandise, including supermarkets, shopping centers, department stores, neighborhood grocers.

6. *Office:* Buildings used for the transaction of business (other than mercantile) accounts or record keeping, and so on. Included are doctors, dentists, libraries, banks, and the like.

7. *Industrial:* Manufacturing, processing, assembling, packaging, repairing of various processes and products, and other activities performed in factories, laboratories, refineries, and similar locations.

8. *Storage:* Storage of goods, merchandise, vehicles, housing of animals, warehouses, garages, freight terminals, grain elevators, and similar structures.

9. *Miscellaneous:* Any other not classified, such as bridges and towers.

CONTENT-HAZARD CLASSIFICATION

The process or material of any given business may offer a hazard peculiar to that particular manufacturing process or material substance. This hazard can be given a broad classification index as a guide to the inherent fire potential.

1. *Low-Hazard Contents:* Noncombustion or low combustion, so that self-propagating fire cannot occur. Evacuation might be necessary because of smoke, fumes, or an outside source of fire.

2. *Ordinary-Hazard Content:* Combustion of moderate rapidity with considerable smoke, but no poisonous fumes or explosion.

3. *High-Hazard Content:* Combustion of extreme rapidity or poisonous fumes or explosion hazard in attendance with the fire.

SURFACE-BURNING TESTS OF BUILDING MATERIALS

The spread of fire across the exposed surface of a combustible material has a direct bearing on the extent and severity of the fire and plays an important role in life safety. Many cases can be cited where loss of life was attributable to the rapid spread of flame and heat over the finished walls and ceilings of the fire building.

Various factors contribute to the flame spread. The thickness and texture of a material, its composition, base materials, and the combustible vapors generated all are instrumental. These characteristics can be tested and classified. Again, standards derived by the NBFU, the NFPA, and the ASTM are the criteria for classification of surface-burning materials.

The medium is the tunnel test. The tunnel is a test chamber 17½ inches wide, 12 inches deep, and 25 feet long. The material to be tested is fastened to the top surface of the tunnel. A gas flame is initiated 8½ feet from one end. The gas flame is so adjusted that a test sample of red oak flooring has a flame spread of 19½ feet in five and one-half minutes. This sample is given a classification of 100. A ¼-inch asbestos cement board is similarly tested and given the classification of 0.

The test specimen is then tested to see whether flame propagates along its entire length, and if so, to determine the time necessary to do so. If the flame does not spread the 19½ feet in ten minutes, the test is stopped and the distance of flame spread is measured. From these measurements a flame-spread classification in relation to the standard of red oak can be determined.

In brief, a flame-spread rating of less than 100 means that flame spread is slower in the test specimen than in red oak; that is, the flame spread is less than 19½ feet in ten minutes.

Other characteristics, such as the amount of smoke generated, and the toxicity, also are measured. The evaluation report will include observations on delamination, shrinking, sagging, and so on.

Under the specifications of this testing method, five classifications for interior-finish materials have been established.[3]

[3]The following classification is given in the NFPA Life Safety Code.

Class A	flame spread	0 – 25
Class B	flame spread	25 – 75
Class C	flame spread	75 – 200
Class D	flame spread	200 – 500
Class E	flame spread	Over 500

Various codes change the class flame-spread rating slightly. However, life safety considerations make it mandatory that all exits and exitways have interior finish of low flame spread—Class A. In other occupancies where people live or gather in large numbers, Class A should also be required. In occupancies that are only partially occupied with smaller groupings of people, Class B or Class C may suffice.

SUMMARY

Broadly speaking, buildings may be grouped in five classifications relating to their fire potential:

1. Fire-resistive
2. Heavy timber
3. Noncombustible
4. Ordinary
5. Wood-frame

Occupancies of buildings and the hazard of their contents can be classified. A correlation can then be made in the building code between the hazard of occupancy and contents and the fire-resistance rating of the structure housing the occupancy.

Occupancy classifications are:

Assembly	Office
Educational	Industrial
Institutional	Storage
Residential	Miscellaneous
Mercantile	

Fire will always tend to travel upward primarily and also to adjacent rooms or combustibles on a horizontal path, particularly where upward travel is impeded.

Fire prevention measures must consider means of blocking off this spread of fire vertically and horizontally. The spread of fire can be slowed primarily by:

1. Limitation of fire areas by
a. Fire doors and fire walls
b. Closing of vertical and horizontal means of communication

The standard time-temperature curve is the guideline used in the fire testing of building construction and materials.

Materials or components of construction that are tested are subjected to controlled temperatures that follow the rate-of-temperature-rise curve to the specific hour for which the materials are being tested. The materials being tested must be under working loads and stresses to be borne in actual field use. Other specialized criteria may be added for certain materials.

The surface burning characteristics of building materials are determined in the tunnel test, a test that gives an indication of the spread of flame propagation over the surface of the material being tested.

A comparison is made with the flame-spread characteristics of red oak and a rating given. A rating less than 100 means that the flame spread of the rated material is slower than that of red oak. A rating over 100 means that the flame propagates faster than it does with red oak.

DISCUSSION TOPICS

1.   List the points on the time-temperature curve that give standard temperatures for the first half-hour of fire.

2.   Describe the criteria for the rating of one-hour fire-resistant material.

3.   Discuss the different building classifications of NFPA No. 220.

4.   A structure is especially vulnerable to the spread of fire in a vertical direction. What are common paths of vertical fire spread?

5.   How can fire be kept from spreading in a horizontal direction?

6.   Discuss the importance of materials with low surface-burning characteristics.

RESEARCH PROJECTS

1.   Canvass the local area and list several buildings for each standard type of construction.

2.   Visit new building sites and examine framing with an eye for possible paths of fire communication.

3.   Locate at least one example of each class of occupancy in buildings in the local area. Evaluate the occupancy and the type of building construction housing the occupancy.

4.   Obtain a copy of the local building code and fire prevention ordinances. Is it a "performance"-type code? Are there sections included by "reference" only?

FURTHER READINGS

Branigan, Francis, *Building Construction for the Fire Service,* 1st ed., National Fire Protection Association, Boston, 1971.

Crane, Theodore, *Architectural Construction,* 2d ed., Wiley, New York, 1967.

Huntington, C. W., *Building Construction,* 3d ed., Wiley, New York, 1963.

Merrith, Frederick, *Building Construction Handbook,* 2d ed., McGraw-Hill, New York, 1965.

The simplistic fire triangle and the more recent fire tetrahedron showed us four ingredients necessary for combustion to take place. Conversely, four methods of extinguishing a fire were described.

1. *Cooling.* If the temperature of a burning material is reduced to below the temperature necessary for ignition to take place, combustion will stop.

2. *Smothering.* If the oxygen source that is necessary for combustion is blocked off, combustion will cease.

3. *Removal of fuel.* If the fuel that is not yet burning is removed, combustion will be stopped when the burning fuel that is left is consumed.

4. *Breaking of the chain reaction.* If an inhibiting agent is introduced that breaks up the chain reaction of combustion, combustion will cease.

The cave dwellers saw fire start, perhaps by lightning, and then saw the rain that followed the lightning put out the fire. Water had always been plentiful, and later they and their followers used it to extinguish their controlled fires. They could not explain what happened, but they knew the fire went out. We now know that the water had *cooled* the burning material to below its ignition temperature and vaporization of flammable gases had ceased.

We can all recall the automatic reflex that has caused someone to throw a blanket or other covering over a sudden flare-up of fire when water was not readily available. Perhaps sand or dirt was used to extinguish the campfire. The fire was literally *smothered,* being "blanketed" from the sustaining oxygen. Yet again, perhaps we were not aware of what actually happened.

EXTINGUISHING AGENTS

We have also seen a burning object snatched from a fire to prevent its further burning. In some cases, we have seen fire "beaten to death" by a stick or club. Both methods are forms of *separating* the burning material from the flame or heat source. (These three methods of extinguishment are illustrated in Figure 8-1.)

A *chain-breaking reaction* is a chemical process that is more sophisticated and not so readily at hand. More sophisticated means are also now available for the control of fire in the three "natural" methods of extinguishment just discussed.

FIRE CLASSIFICATION

A study of extinguishing agents presupposes a knowledge of different types or categories of fires. The burning substance, although a combustible, may vary considerably from other burning

Figure 8-1   Extinguishment of fire by natural means—cooling, smothering, and the removal of fuel. The woman on the right is helping to put out the fire by cooling. The center man is attempting to extinguish the fire by smothering it with sand. The man on the left is removing some of the burning fuel.

materials. One may be a liquid, another a solid, a third a gas. Other factors may also be present.

To differentiate, fires have been grouped into four general classifications. These classifications, and appropriate extinguishing agents, are listed in Figure 8-2.

Class A Fire

A class A fire is the type of fire that occurs in ordinary combustible materials—wood, paper, animal and vegetable fibers. These are the organic chemical compounds that contain carbon along with some hydrogen, oxygen, and nitrogen. Most fires come within this classification. They are normally controlled and extinguished by cooling or the removal of fuel.

Class B Fire

That special class of fire that is encountered in the flammable liquids and gases is designated as class B. Class B fires may be controlled by cooling. The procedure consists mainly of cooling the tanks or containers of flammable liquids to prevent excess pressures or vaporizing. Extinguishment is through smothering, preventing oxygen from reaching the seat of the fire; occasionally by

Figure 8-2 Classification of fires and extinguishing agents.

FIRE SUBSTANCE	EXTINGUISHING AGENT	WATER	CO_2	FOAM	DRY CHEMICAL	DRY POWDER
Class A	Ordinary Combustibles	Yes	Small Surface Fires Only			No
Class B	Flammable Liquids	Limited use Water Fog	Yes	Yes	Yes	No
Class C	Electrical Equipment	No	Yes	No	Yes	No
Class D	Combustible Metals	No	No	No	No	Special Dry Powders

the removal of fuel; or by cooling in those flammable liquids that are capable of being miscible (or mixable) with water. The addition of water to miscible flammable liquids not only cools the liquid, but also raises the flash point of the liquid by dilution.

Class C Fire

Fires occurring in electically energized equipment are grouped in class C. Fires involving electrical equipment must be fought with a nonconducting extinguishing agent. Control is obtained by smothering—the blocking off of the oxygen supply to the fire. Complete extinguishment may require the use of a cooling agent, probably water, after the equipment has been deenergized.

Class D Fire

The combustible metals and some highly reactive flammable liquids are placed within this category—the one class of fire that seems to defy all attempts at extinguishment by normal means. In many cases, there are greater flare-ups, even the danger of explosion when accepted extinguishing agents are used. Frequently the only means of control and extinguishment is by smothering with special powders usually kept on the premises.

Class A, B, and C fires are normally controlled and extinguished by standard fire-extinguishing agents. These agents are in general use by fire departments and industries concerned with fire extinguishment. Despite the advances in technological knowledge and the contributions of natural scientists to fire extinguishment by various means, water is still the most widely used and most efficient extinguishing agent.

WATER

Water is relatively cheap, and it is much more easily procurable and available than are any other known extinguishants. Some of the physical properties that make it our best extinguishing agent should be understood.

1. At ordinary temperatures, water is a stable liquid and is not easily decomposed. (Some of the combustible metals do have the

potential of decomposing water into its components of hydrogen and oxygen and then using the released oxygen for support of their own combustion.)

2. Specific heat is defined as the number of Btu needed to raise the temperature of one pound of a substance 1°F. In the case of water, one Btu is required. No other substance has as high a specific heat. (Kerosene, for example, has a specific heat of .50.) This means that more Btu, or units of heat, will be required to raise the temperature of water, say 100°F, than will be needed to raise the temperature of any other substance the same amount. The latent heat of vaporization of water presents an even higher heat-absorbing capacity. To convert one pound of water at 212°F to steam at 212°F requires 970.3 Btu. This is the latent heat of vaporization. To convert one pound of water at 62°F to 212°F requires 150 Btu. To convert water at 212°F to steam at 212°F requires 970.3 Btu. Total Btu absorbed per pound of water in this process is 1,120.3. In this lies the potential of water for fire extinguishment. The most efficient means of fire extinguishment is the removal of heat from the combustible material. Heat, we know, is a form of energy measured in Btu. No other extinguishant compares with water in heat-absorbing capability. (See Figure 8-3.)

3. In the process of converting from a liquid at 212°F to steam at 212°F, water increases in volume 1,700 times—the expansion is so rapid that the steam will tend to replace smoke, heat, and fire gases as well as the air needed for combustion. The steam then will tend to help smother the fire by reducing available oxygen.

4. Water may dilute certain flammable materials that are water-soluble to a point where extinguishment is effected. Thick, viscous liquids, such as heavy fuel oil, will tend to froth up when struck with water. This emulsification also can aid extinguishment.

There are limitations and some hazards to the use of water as an extinguishing agent. Its ability to conduct electricity poses the problem of fire in electrical equipment and the use of water. The reactivity of water with materials such as magnesium and sodium also precludes its use in fires involving these chemicals. Sodium and other elements that are in the same periodic grouping as sodium (group 1), and magnesium and other elements in group 2, are particularly hazardous when struck with water. The highly flammable gas hydrogen is released in the reactive process and ignition or explosion can result.

Figure 8-3 High-caliber streams in use on the fire ground. Because of the extent of fire, it is impossible to obtain the total cooling effectiveness of the water, and much of it will merely run off.

Courtesy of the Chicago Fire Department.

But even in the ordinary class A fire, there are limitations to the effectiveness of water. Water in its liquid state maintains a certain cohesion of its own that retards its deep penetration into most materials. The viscosity of water, however, is such that it will tend to run off rather than remain as a cover and a smothering agent on the surface of the burning material.

To overcome these handicaps of water, chemical agents can be added to change its basic characteristics. Wetting agents can be added that will reduce the surface-tension characteristic so that it will flow more readily and penetrate more deeply into piles of combustibles that are packed, baled, or stacked. Wetting agents can be also put to good advantage in the overhaul and final extinguishment of stubborn, deep-seated fires. Wetting-agent solutions may be combined with foam for oxygen exclusion and also for penetration resulting in heat absorption.

Where there is a need for "thicker" water to prevent runoff, chemicals can be added that will aid water in forming a thicker blanketing film. This thicker layer can absorb more heat and also act as a smothering agent. Thicker water plays an important role in forest firefighting.

Aqueous film-forming foam (AFFF) is a more recent example of what can be accomplished with water combined with an additive. The additive in this case consists of active fluorine chemicals that, in solution with water, "float" on top of the flammable liquid. Water released from the foam bubbles does not sink, but spreads over the surface and aids in cooling the burning fuel. Very fast knockdown of flammable liquid fires has been accomplished more rapidly with this film-forming foam than with conventional foam. The additive is compatible with the bicarbonate-base dry chemicals and is used as a twinned attack team on flammable liquids, aircraft crash fires in particular. A unit with dual containers, one of potassium bicarbonate dry chemical and the other of "light" water, has separate hoses and nozzles in each container that form an integral unit that can be led out together. Each nozzle can be triggered separately or simultaneously. The liquid has been called "light" water because of its characteristics.

"Slippery" water is another newly developed water. A special chemical dissolved in water reduces friction loss in hose lines to such an extent that water flowage is increased 70 percent. Nozzle reach is said to be doubled. The fire department can therefore use a smaller-diameter, more maneuverable hose than the standard sizes now used—and obtain equal flow. New York City is experimenting both with a pumper carrying this additive and with 1¾-inch hose replacing the conventional 2½-inch hose.

CARBON DIOXIDE

Carbon dioxide has a long history as an extinguishing agent, being used primarily for flammable liquid fires and electrical equipment fires. Carbon dioxide, CO_2, is noncombustible and does not react with most substances. It is a gas, but it can be easily liquefied under pressure and is normally stored under such a condition. It provides its own pressure for release and blankets the fire area when released in sufficient amounts. Being heavier than air, it replaces the air above the fire and smothers it. CO_2 is mildly toxic, as stated in the discussion of fire gases in Chapter 6. It can induce unconsciousness in the person trapped in a 9 percent concentration.

Carbon dioxide is a nonconductor of electricity, and it is this attribute that makes it a valuable aid in controlling the electrically energized equipment fire.

The principal property of extinguishment possessed by CO_2 is its smothering, or air-replacement, capability. One pound of liquid CO_2 will vaporize to 8 cubic feet of gas and displace an equal volume of air. It does possess some cooling properties, but they are minor when compared with those of water. When released as a liquid, CO_2 expands to a gas so rapidly that much of it is converted into a "snow" at a temperature of $-110°F$. As this "snow" changes to a gas, some heat is absorbed from the fire area and some cooling results. The latent heat of CO_2 "snow" is 246.9 Btu per pound. Compare this heat with the latent heat of one pound of water—1,120 Btu from water at $32°F$ to steam at $212°F$.

There are some disadvantages and precautions to be considered in the use of CO_2 as an extinguishant. There is a possibility of a rekindle from hot embers after the carbon dioxide gas has been dissipated and no longer affords a smothering effect. This is the reason that water may have to be used on the deep-seated electrical fire after the power current has been shut off.

Carbon dioxide has been used successfully on flammable liquid fires and, in particular, on air-crash fires where a quick knockdown is needed. Here, too, the possibility of rekindle must be considered. CO_2 cannot be used successfully in those chemical fires, such as cellulose nitrate, that generate their own oxygen supply. The same, of course, is true of any extinguishant that utilizes a smothering action. Certain metals, such as sodium, potassium, magnesium,

Figure 8-4 A carbon dioxide extinguisher in use at a hospital staff training exercise.

Courtesy of Fredriksen & Sons Fire Equipment Company, Inc.

titanium, zirconium, and the metal hydrides, are highly reactive with carbon dioxide and will decompose it, negating the smothering action. (Figure 8-4 shows the use of a carbon dioxide extinguisher in a hospital.)

FOAM

Foam is considered to be the best extinguishant for flammable liquid fires that cover a large area. (See Figure 8-5.) Because of its action of blanketing the fire area, it can be flowed on in progressive steps until total extinguishment is obtained. The coherent blanket formed extinguishes by smothering. The water content of the foam extinguishes by cooling. Foam must float on top of a flammable liquid to be effective as a smothering agent. In those cases where the blanket cover is broken, only small patches of fire will result which can be recovered to effect extinguishment again. Care must be exercised to be sure that the blanket is not widely broken by a water stream or an incompatible dry chemical. This requirement predicates a foam that is stable in its basic formula.

There are two common types of foam, chemical and mechanical.

Chemical foam is generated by the chemical reaction of two

Figure 8-5 A foam extinguishing agent in use on a flammable liquid fire.

chemicals in water. One of the two chemicals is an alkaline salt solution, such as aluminum sulfate; the second is usually sodium bicarbonate with an added foaming agent and a stabilizer. The chemicals can be added to the water from two separate sources, or the chemicals may be premixed in a two-in-one container.

Mechanical foam (air foam) is generated by introducing a small proportion of the foaming agent into water and then mixing it turbulently with air. The degree of mixing will determine the characteristics of the foam blanket—watery or viscous. Water temperature is not as serious a factor as it is in the chemical foam. The foaming agent can be a protein base or a synthetic type. The liquid concentrate is normally found in two concentrations—3 percent or 6 percent. The 3 percent is mixed with 97 percent water, and the 6 percent with 94 percent water. This foam has a low expansion ratio, and expansion may be from 6:1 to 10:1.

The foam blanket prevents formation of ignitable vapors, offers protection against heat radiation, retards toxic gas spread, and blocks oxygen from reaching the burning liquids. The water in the foam has a cooling action. Through a careful application of water spray, water in the foam lost to the heat of the fire can be replaced, but this must be done carefully to avoid the possibility of dissipating the foam blanket. Mention has already been made of the double benefits of a compatible foam dry chemical and foam used in multiple attack.

Because of the water content, foam cannot be used on electrical fires or on those chemicals that react with water.

High-Expansion Foam

A method has long been sought to extinguish those fires beneath grade level where ventilation is difficult. Research in England for an extinguishant for mine fires brought on development of high-expansion foam, a detergent product.

Expansion rate is over 100:1, in most instances 1,000:1. High-expansion foam is nothing more than bubbles mechanically generated by the passage of a large volume of air through a net or screen which has been wetted by an aqueous solution of the foaming detergent. The generated bubbles are blown through a large-diameter collapsible tube to the fire area, and they quickly achieve total flooding of any confined space. (Figure 8-6 illustrates such a use.) Vapors, heat, and smoke are displaced; heat radiation and transmission are blocked. The water in the foam, here too, absorbs some of the heat of the fire.

Total flooding may be blocked by partitions that stop the flow to the desired area. In other instances, open doors or windows permit the foam to flow from the fire area. It can be used above grade as well as in subgrade areas. It is important that the necessary air for foam generation is not drawn from the fire building: The heated air, smoke, and gases of the fire will cause a sharp reduction in the volume and possible physical disruption of the blanket.

Figure 8-6   High-expansion foam being used in an attempt at total flooding of a confined fire area.

Courtesy of National Fire Company.

It is possible for firefighters to walk about in the foam-filled area, but there is also the possibility of their becoming "lost" and disoriented. Backup water lines may have to be used if foam is dissipated before total cooling and extinguishment are achieved.

DRY CHEMICALS

Dry chemical extinguishing agents are dry chemicals in a finely ground state, varying somewhat according to the purpose for which they are intended. Dry chemicals used in the extinguishment of class B flammable liquids and class C electrical fires were originally of sodium bicarbonate with additives, such as metallic stearates or silicones. Both sodium and potassium are in group 1 of the periodic table and have one electron in the outer orbit. Potassium, however, is more active than sodium. After success was attained with sodium mixtures as an extinguishing agent, experiments were conducted using potassium mixtures as the agent. Because of the greater reactivity, potassium proved to be a better extinguishant than sodium on the class B and C fires.

A multipurpose dry chemical extinguishant is also in use. Ammonium phosphate is the base powder with other additives mixed in. The multipurpose dry chemical can be used on class A as well as class B and C fires. On class A fires, the chemical melts upon striking the hot surface, becomes sticky, and adheres to the material.

Dry chemicals are stable at normal or low temperatures. A temperature of 140°F is said to be at the top limit for storage purposes.

The dry chemical requires an expelling agent to propel it from its container. The pressure of a nonflammable gas expels the dry chemical, rapidly filling the discharge area with the chemical. The fine particles are nontoxic, although visibility is impaired and breathing is very difficult within the discharge area.

Extinguishment was at first thought to be caused by smothering and some cooling. The released CO_2, it was believed, aided in extinguishment by a cooling process. However, tests revealed that 5 pounds of chemical was as effective in extinguishment as 10 pounds of CO_2. This finding stimulated more research on the basic fire-triangle theory of extinguishment and the addition of a fourth side to the fire triangle—the chain-breaking reaction hypothesis. The free radicals of combustion are prevented from reacting with one another by the reaction with the dry chemical. The chain of combustion is broken.

There are limitations to the use of dry chemicals. When expelled, they dissipate, and oxygen is again available. The fire can rekindle if an ignition source or hot embers remain. While they are effective in electrical fires, dry chemicals can be harmful to electrical contacts and relays. Also, they are not effective for fires involving materials that generate their own oxygen supply.

Dry chemicals do prove particularly effective on flammable liquid fires (Figure 8-7) and especially on aircraft crash fires because of the quick extinguishment attained. The inherent danger is the rekindle possibility when the dry chemical dissipates. To overcome this hazard, foam can be used in conjunction with the dry chemical: the chemical to achieve initial fast extinguishment, and the foam to control and maintain extinguishment. During initial experimentation in this method, the dry chemical proved incompatible with mechanical, air foam, breaking the foam cover down and destroying its effectiveness. A foam-compatible dry chemical was then

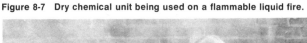
Figure 8-7 Dry chemical unit being used on a flammable liquid fire.

formulated. In conjunction with foam, it has proven particularly effective for aircraft crash fires. (Refer to the section on "light" water and its compatibility with potassium-base dry chemicals.)

HALOGENATED AGENTS

A halogenated extinguishing agent is formed by the introduction of a halogen into an organic compound (carbon). The halogens, the chemicals in group 7 of the periodic table of the elements, are chlorine, fluorine, bromine, iodine, and astatine. One of the most successful of the halogenated agents is bromotrifluoromethane— (trade names are Halon 1301 or Dupont felon fe 1301). The chemical formula is $CBrF_3$.

Extinguishment appears to be obtained by a chemical chain-breaking reaction which stops the combustion process. The chemical reaction theory is as follows: A bromide radical, Br, in the agent reacts with the fuel. In this chemical reaction, hydrogen bromide, HBr, is one substance formed.

$$R - H + Br \rightarrow R + HBr$$

The HBr in turn reacts with the hydroxyl OH radical found in the normal combustion process. Water vapor and bromine, Br, is the end product.

$$OH + HBr \rightarrow H_2O + Br \text{ radical}$$

This free bromine radical reacts with more fuel to remove hydrogen to form HBr, which, again in turn, reacts with the OH radical. Thus, the chain reaction of combustion is broken by the removal of H–OH and O. The halogens are toxic, but the concentrations encountered are more irritating than lethal.

DRY POWDERS

Magnesium, titanium, sodium, potassium, and several other metals, when burning, are capable of generating their own oxygen supply. Fires in these metals, known as class D fires, are very difficult to extinguish. All are very reactive with water, and little success is obtained with other generally accepted extinguishing agents.

Special extinguishing agents, mostly dry powders, are used for these specific fire hazards. While chemical in themselves, these agents are classified as dry "powders" to differentiate them from the dry chemicals used on most class B and C fires, but ineffective on these fires. These special dry powders are not effective on all of the combustible metals, some powders being limited to a specific metal. Control, rather than extinguishment, is the best that can be obtained with some powders.

In some cases, the powder, or liquid, is applied by means of a standard-type extinguisher. In other cases of class D fires, the powder may be shoveled over the burning metals. These means of control or extinguishment for the particular combustible metal will normally be found stored on the plant site rather than carried as a standard piece of equipment by the responding fire department.

G-1 Powder, a trade-name extinguishant, is a mixture of granulated graphite and an organic phosphate that must be shoveled onto the burning mass. The organic phosphates vaporize and tend to exclude air. The graphite offers some smothering effect plus removal of much of the heat.

Met-L-X Powder can be used in an extinguisher with a gas as the propellant. Additives to a sodium chloride–base formula make it free-flowing and capable of being expelled from the extinguisher. In use, the heat of the fire causes a solidification of the fine particles into a unified mass that smothers the fire.

Both G-1 and Met-L-X are nontoxic and noncombustible. While there is some variance in degree of effectiveness, both can be classified as controlling or extinguishing agents of magnesium, sodium, potassium, titanium, lithium, and zirconium.

TMB liquid (trimethoxyboroxine) can also be used in an extinguisher as a controlling or an extinguishing agent on magnesium, zirconium, or titanium fires. It is a flammable liquid in itself.

It has been observed that burning magnesium is in one sense two fires. The first is a fire of vapors that burns quite hot, over 5000°F and with a white-hot flame. The second fire is a cooler and slower fire on the metal's surface. TMB liquid controls the white-hot fire by retarding vaporization in a smothering action. The TMB liquid first ignites in its own fire, then forms a smothering coat of boron oxide. Once the white-hot fire is controlled, water spray is capable of controlling the second, cooler surface fire without the violent reaction normally encountered when water is used on a magnesium fire. This technique is being used on aircraft fires.

Other industries using combustible metals also quite frequently have other powders that can be shoveled or thrown on the burning metal for control, if not extinguishment. Some of these agents are foundry flux, asbestos, talc or graphite powder, and soda ash. The user of the material has the responsibility of having some means of control for that particular combustible metal at hand.

Accepted standard extinguishing agents are now in practical use in various ways. Most familiar to the average individual are the portable extinguishers kept in various occupancies as the first line of defense against fire. Fire code requirements list the type and number deemed necessary for the various structural and occupancy factors at the specific site. Sizes of the various extinguishers vary, depending on expected use and type of extinguishant. They range from a 2½-gallon water pump through a 40-gallon wheeled foam extinguisher to a wheeled 750-pound refrigerated carbon dioxide unit. In the selection of the proper type and size of extinguisher, consideration is given to the specific combustible hazard and the expected severity of the fire potential. These conditions are then matched with the characteristics and capability of the extinguishing agents.

The modern, well-equipped fire department will carry, or have access to, agents suitable for the average fire it will encounter. Class D fires can present special problems. The property owner must assume added responsibility for fire protection when this class of fire is possible. The quantities of agents carried for class B and C fires may be limited, but the fire ground commander should have a backup supply available. He should have a dry chemical unit and a foam-generating unit at hand. Proper airport protection requires a foam crash unit or a twinned "light" water and dry chemical unit.

Building codes, insurance underwriters, and just good common sense dictate that certain occupancies and hazards will have fixed fire protection systems that can be automatically activated and placed in service before portable systems can be put in use or the fire service can respond. A common first line of defense is a sprinkler system. Other systems use carbon dioxide, dry chemical, foam, or the halogenated agents. Again, hazard and severity determine the type and method of protection. Some representative installations are:

1. Carbon dioxide systems for surface flammable liquids; for equipment where damage from the extinguishing agent must be minimized; and for protection of electrical installations

2. Dry chemical systems for flammable liquids and electrical installations

3. Foam systems for flammable liquids where reignition is possible, for example, large outdoor storage tanks for gasoline

4. Halogenated agents as protection against flammable liquid fire potentials; in particular, used by the aircraft industry as an aircraft-engine fire extinguishant in a fixed installation

(Chapter 13 contains additional information on fire protection systems.)

SUMMARY

Fire extinguishment can take place by elimination of one of the four known factors involved in combustion. The basic methods of doing this are:

1. Cooling: Reduce the temperature of the burning substance below its ignition temperature.

2. Smothering: Prevent the source of oxygen necessary for combustion to take place from reaching the seat of the fire.

3. Removal of fuel: Remove substances not yet ignited from the fire area.

4. Rupture of chain reaction: Introduce another substance into the fire area that will react chemically with products of combustion and break up the chain reaction feeding combustion.

Because of the varying nature of the fire itself, fires have been grouped into four general classes:

Class A—Ordinary combustible materials

Class B—Flammable liquids and gases

Class C—Fires involving energized electrical equipment

Class D—Combustible metals

There are various extinguishing agents available, many suited for a particular classification of fire that requires its own method of extinguishment.

Water is by far the cheapest, most plentiful, and best extinguishing agent available. Its extinguishing power lies in its ability to cool by absorbing the heat energy of the fire. Water converted to steam by the generated heat not only has cooled the fire, but the steam tends to smother the fire by blanketing out oxygen.

Certain limitations to water—its inability to penetrate readily and the fact it runs off a surface quickly—can be overcome by specific additives. Its reactivity with combustible metals and its conductivity of electricity forestalls its use in fires involving these substances.

Carbon dioxide, CO_2, is a gas that liquefies under pressure and extinguishes by smothering and slight cooling. As a nonconductor of electricity, it is useful on the class C type of fire, and also in smothering the class B fire.

Foam extinguishing agents are of several types. Chemical, mechanical, and high-expansion foams are examples. Chemical foam is the reaction of dry chemicals and water; mechanical foam, the mixing of a liquid foaming agent, water, and air; high-expansion foam is a detergent type that creates mechanically generated bubbles by the passage of air through a net or screen holding the solution. Foam is useful on flammable liquid fires, as it blankets the area and separates the fire from the oxygen source. Some cooling also takes place. High-expansion foam can be used for total flooding of a confined fire area.

Dry chemicals in a finely ground state, expelled by a gas, nitrogen, or CO_2, blanket the area with a fine powder and extinguish by smothering. There is no cooling effect and there is a danger of rekindle from heated embers when the chemical dissipates. The chemicals also react in a chain-breaking reaction to retard combustion, which is useful on class B and class C fires.

Halogenated agents contain chlorine, bromine, or fluorine, the group 7 elements of the periodic table. Extinguishment is by the chain-breaking reaction. Some toxic gases are released, so the principal use is in fixed installations.

Dry powders are the special powders used in the class D combustible metal fires. Some specific trade names are G-1 Powder, Met-L-X, and TMB liquid. These extinguishing agents are used in portable extinguishers or in fixed protection systems that are automatically activated when fire occurs.

DISCUSSION TOPICS

1. Match each class of fire with the type of fire it represents.

2. List some characteristics of water that make it the ideal fire-extinguishing agent.

3. Select an extinguishing agent that can be used for each class of fire.

4. What are some handicaps of water as an extinguishing agent?

5. List types and uses of foam in firefighting.

6. What distinguishes a dry powder from a dry chemical as a fire-extinguishing agent?

RESEARCH PROJECTS

1. There are four general classifications of fire. List several locations in the community that could be subject to a fire in each specific classification.

2. List at least two extinguishing agents that can be used at each of the locations you have listed.

3. Visit a location using a fixed extinguishing system other than water.

4. Set up a demonstration using several different classes of portable extinguishers.

5. Outline the characteristics of water that make it a good extinguishing agent. Add a list of the detrimental features of water when used for fire extinguishment. Describe the means of overcoming some of these handicaps in the use of water.

FURTHER READINGS

Casey, James F. (ed.), *Fire Service Hydraulics,* 2d ed., Reuben H. Donnelley Corp., New York, 1970.

Tryon, George H. (ed.), *Fire Protection Handbook*, 13th ed., National Fire Protection Association, Boston, 1969.

Pamphlets and brochures obtainable from manufacturers of fire-extinguishing equipment.

The craftsman is only as good as the tools he must work with. The firefighter, as a craftsman skilled in the extinguishment of fire, can only perform his duties well if he has the proper tools and equipment.

The technology that has improved the life style of the twentieth century has also advanced the art of fire extinguishment through more effective fire apparatus and power tools. But there is room for improvement in both apparatus equipment and hand tools. Part of the problem is the old law of supply and demand. If the demand were greater, the supply, and the innovative use, of fire department equipment would be greater. The fire service needs specialized equipment. Its motor-driven apparatus necessitates strict performance requirements because of the emergency service performed. Yet the demand is not large enough for a manufacturer to go all out in developing production techniques and stockpiling spare parts. Each piece of motorized equipment, built to the specifications of the purchasing fire department, in a sense is hand-assembled and hence costly. Small hand tools are also, in many instances, not mass-produced.

THE PUMPER ENGINE AND FIRE PUMPS

The first line of attack and defense in firefighting is the pumper or fire engine. The pumper carries a water pump that is capable of taking water from a hydrant or open body of water and sending it through hose lines to the attack point of the fire. Sufficient water pressure must be pumped so that the water can be delivered to the seat of the fire. The "heart" of the fire pumping engine is the fire pump.

FIRE DEPARTMENT EQUIPMENT

9

In the early 1950s, 750-gpm pumpers appeared to be the accepted standard for fire departments. The trend today seems to be to standardize around a 1,000-gpm pumper, with larger municipalities having some 2,000-gpm pumpers. (Figure 9-4 shows two 1,000-gpm pumpers.)

This is not to imply that the 1,000-gpm pumper is always delivering 1,000 gpm of water on the fire building. A review of some simple principles of physics and an examination of the typical fire pump will reveal why.

The ability of a fire pumper to do effective work is measured in water horsepower—in other words, in terms of pressure in pounds per square inch and gallons per minute. This is similar to the standard horsepower.

After standards of measurement have been determined, it is necessary to define standards of performance acceptable to the fire service. NFPA Standard No. 19 (see the reading list at the end of this chapter) delineates standards of performance for fire pumps. These are the accepted standards for fire department pumpers. They state that a fire pump must be capable of delivering its rated capacity — gallons per minute (gpm) at 150 pounds per square inch (psi).

These two requirements, gpm and 150 psi, multiplied together, give the required work capacity of the fire pump. It follows, then, that if greater pressure is needed to deliver water to the fire building, the increase in pressure will cause a decrease in gpm delivered. Because of additional loss due to increased slippage and friction losses in the pump itself when working at higher pressure, the pumper standards accept increased gallonage loss. The rated capacities that must be met are:

Rated capacity at 150-pound pressure

Seven-tenths of rated capacity at 200-pound pressure

One-half of rated capacity at 250-pound pressure

For a "standard" 1,000-gpm pumper, the ratings are:

1,000 gpm at 150-pound pressure

700 gpm at 200-pound pressure

500 gpm at 250-pound pressure

Early types of fire pumps in use on fire department engines were the positive-displacement pumps—the piston pump and the rotary gear pump.

Piston Pump

The piston pump was the first pump used in the fire service. It was this type of pump that saw use in the early hand water tubs. This is a positive-displacement type of pump that can be "simple-acting," that is, it discharges water when the piston is cycling in one direction only; or it may be "double-acting," capable of discharging water also in the reverse cycle. With each stroke, a definite volume of water is discharged which must have an open flow out or the piston will become "bottled up." (Figure 9-1 illustrates this type of pump.)

Rotary Gear Pump

The use of rotary fire pumps began in the era of horse-drawn steamer pumpers. It too is a positive-displacement pump. It can be

Figure 9-1   A double-acting piston pump. (a) As the piston moves to the left, the water is compressed and pressurized. This forces open gate B and pumps the water out under pressure to the discharge side. Gate D opens and permits more water to enter the cylinder chamber from the intake source. Gates A and C remain closed. In (b), as the piston reverses its stroke, gate C is opened, permitting water to be discharged. Gate A opens on the intake side. Gates B and D close.

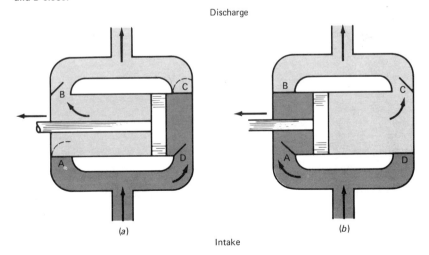

one of two types; one with gear-type rotors and the other with lobe rotors (shown in Figure 9-2). In either type the rotors turn in opposite directions, forcing a given amount of water from the suction chamber to the discharge manifold.

Centrifugal Pump

The piston and rotary pumps have been largely replaced by the centrifugal fire pump. This pump consists of one or more rotating impellers mounted in a volute-shaped discharge chamber. Water enters the impeller at its eye, and is picked up by the rotating impeller; centrifugal force gives energy and pressure to the water as it enters the volute chamber. In the volute, the velocity of the flow is converted to pressure energy as the water passes to the discharge chamber. Each impeller is termed a stage. A centrifugal pump may be a single-stage or a multistage parallel-series operation.

The average fire pumper will consist of a fire pump with two stages that can be operated in either a parallel or a series arrangement. (See Figure 9-3.) In the parallel setting, each impeller takes water from the source and discharges it to the hose line. The total volume of water is the total of the two stages. The pumper is said to be operating at capacity with each stage paralleling the other. In a 1,000-gpm pumper, for example, each stage contributes a potential 500 gpm to the flow to the fire building.

Figure 9-2 A lobe rotor type of rotary pump. Water enters the pump at the intake port, is carried in spaces between the gear teeth or lobes and casing, and is forced out through the discharge port.

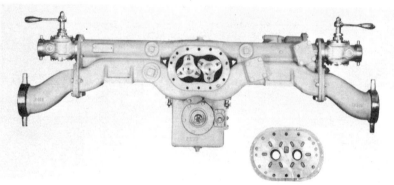

Courtesy of Waterous Company, St. Paul, Minnesota.

When a fluid, such as water, flows through a constraining conduit, say a pipe or hose, pressure readings taken at both ends indicate a change in pressure. When both readings are taken at the same elevation, the pressure will drop as the fluid flows from point A to point B. This loss in pressure is known as friction loss.

Figure 9-3 Schematic of a two-stage parallel-series centrifugal pump. Top view: Parallel operation, or volume. Bottom view: Series operation, or pressure.

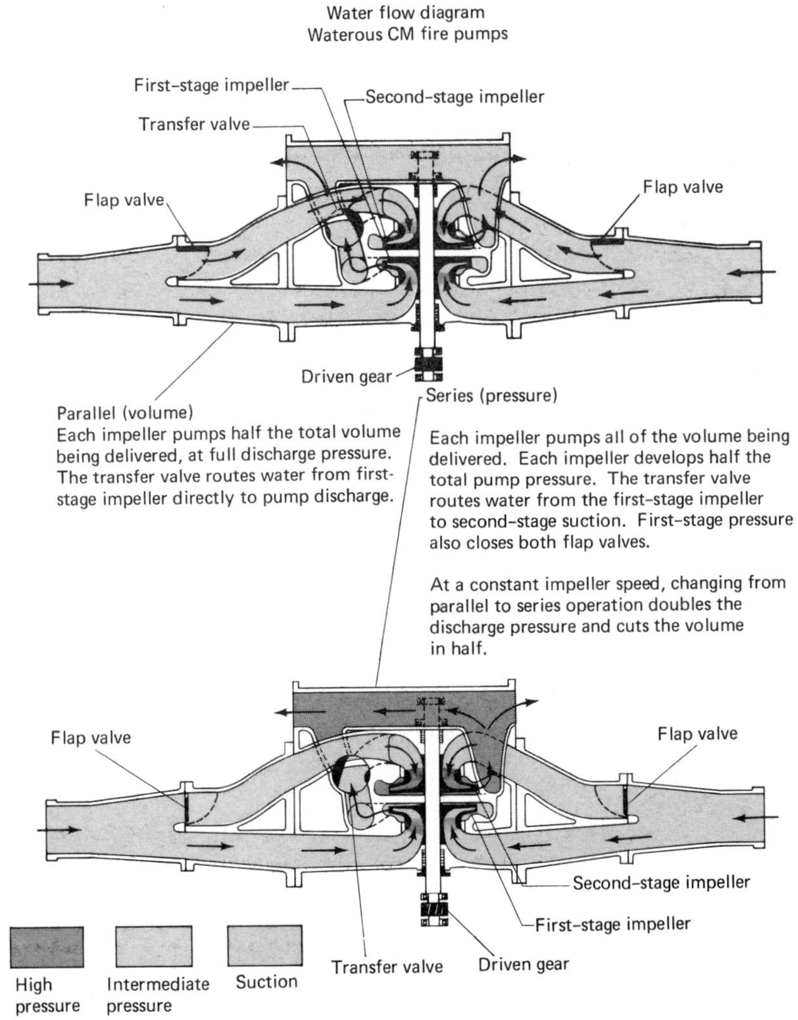

Water flow diagram
Waterous CM fire pumps

First-stage impeller
Second-stage impeller
Transfer valve
Flap valve
Flap valve
Driven gear

Parallel (volume)
Each impeller pumps half the total volume being delivered, at full discharge pressure. The transfer valve routes water from first-stage impeller directly to pump discharge.

Series (pressure)
Each impeller pumps all of the volume being delivered. Each impeller develops half the total pump pressure. The transfer valve routes water from the first-stage impeller to second-stage suction. First-stage pressure also closes both flap valves.

At a constant impeller speed, changing from parallel to series operation doubles the discharge pressure and cuts the volume in half.

Flap valve
Flap valve
Second-stage impeller
First-stage impeller
Transfer valve
Driven gear

High pressure   Intermediate pressure   Suction

Courtesy of Waterous Company, St. Paul, Minnesota.

Further friction loss is encountered in the condition of hose lining (smoothness, age, and the like) and the straightness of hose lead-out (bends and kinks in hose lead-out cause additional friction loss). If the water is being pumped to a higher elevation, pressure is lost overcoming the force of gravity; conversely, pressure is gained when water is pumped to a lower elevation. Additional friction loss results from the intricate piping system of standpipes and sprinklers.

Much of the engine pressure is used to overcome this friction loss in the hose lead-out. As the distance from pump to fire scene increases, pressure loss increases in direct proportion to the length of hose lead-out. Pressure at the pump must be raised to overcome this friction loss. As stated earlier, this greater pressure necessitates a sacrifice of gpm available. As distance to the fire building increases, the limit of a pumper to supply pressure may be reached. Then, if further pressure is needed, a change in the operation of the two stages must be made. This results in a *series*, or pressure, setup. In this setup, stage 1 passes on its capacity, a potential 500 gpm (again, in the case of a 1,000-gpm pumper), to stage 2. The total gallonage of water, 500 gallons, cannot now be increased into the outlet leading to the fire building, but the pressure can be doubled. In effect, the 1,000-gpm pumper has now become a 500-gallon-per-minute pumper, but the ability to send this water under pressure has nearly doubled.

TWO-STAGE PARALLEL-SERIES PUMP

The most commonly accepted fire pump is a two-stage parallel-series type of centrifugal pump. Operated in parallel, it can deliver its maximum capacity at lower pressures. In series, it can deliver higher pressures where needed, but in reduced volume.

SINGLE-STAGE PUMP

A single-stage centrifugal pump is used where water needs are limited and pumping distances are not overly long. Most commonly, it has a capacity of 250 or 500 gallons per minute, although a 750-gpm single-stage pump does exist. With increased pumping distance, the water-pumping capacity must drop. Consequently, the maximum feasible engine revolutions per minute (rpm) and pumping pressures needed to overcome friction loss are soon reached.

TWO-STAGE SERIES PUMP

A two-stage series pump can be used where the pumping distance is great and extra pump pressure is needed. Each impeller stage can then supply half the necessary pressure, permitting the engine to operate at lower rpm and the capacity to be maintained closer to rated capacity.

THREE-STAGE PARALLEL-SERIES PUMP

Where greater pressures are needed than those obtainable in a two-stage series or a two-stage parallel-series hook-up, this type of pump can be used. In essence, this three-stage pump is a two-stage parallel-series pump with a third stage added to increase pressure where pressure over 300 psi is needed.

FOUR-STAGE PARALLEL-SERIES PUMP

This type of pump can be used to discharge water in three different settings.

1. Each stage discharges its volume to the discharge port. The pump is operating in parallel and discharging its rated capacity at 150 psi.

2. The four stages operate in pairs of two. Stages 1 and 2 are in series and stages 3 and 4 are in series. As pairs, they can operate as a parallel or capacity pump, each pair discharging its volume of water into the discharge port.

3. In the series setting, water enters the pump through the intake source only at stage 1. The water is passed on from stage to stage where the pressure is increased but the volume must remain constant. The increased pressure in the four-stage series position enables water to be sent long distances or to high elevations. A four-stage series-only pump is also used on high-pressure fog apparatus where low volume at high pressure is desired.

There is a minor handicap to the operation of the centrifugal type of pump, but it is not insurmountable. If it is hooked up to a supply of water under pressure, such as a hydrant, the centrifugal type of pump operates immediately. However, unlike the positive-displacement types of pumps, which are self-priming, the centrifugal pump cannot create its own suction to draft water from a standing body of water. "Suction" may be a misnomer. In actuality, the pump must be capable of removing some of the air from the suction hose. Once the atmospheric pressure in the hose becomes less than the outside atmospheric pressure, the outside pressure will force water into the hose to displace the "vacuum"; the pump is said to have "lifted" or "drafted" water. The centrifugal pump cannot do this. It must have the impellers primed with water and all air displaced. These requirements can be met by a small rotary pump or some other means that can create a vacuum in the suction line so that normal atmospheric pressure can then fill the suction line with water to the impellers.

Water Tanks and Tankers

Suburban and rural fire departments often face the problems of lack of available water and minimal need for a ladder truck. In these departments, the pumper can be supplied with additional ladders and also carry a water tank of 300- to 1,000-gallon capacity. The apparatus then becomes a "combination."

In addition to this integral source of supply, a tanker capable of carrying 1,000 to 1,500 gallons of water can be part of the attack force. A shuttle relay can be set up to deliver water to the fire scene from the nearest source. These tankers usually carry an auxiliary pump for filling their own tanks and for dumping their supply into the tank of a pumper or a portable storage tank at the fire scene.

The urban fire department may also take advantage of a 300- to 500-gallon water tank as a component of the fire pumper. In large metropolitan city areas, such as expressways and industrial districts divided by many railroad tracks where water can only be supplied by long lead-outs or relay pumping, and even in those areas where hydrants are readily available, there is much to be said for the "instant" water offered from a water tank carried on the responding pumper.

The fire engine, or pumper, will carry a complement of hose to deliver water discharged from its pump to the seat of the fire.

Fire Hose

Individual department standards will vary as to amount and size of hose to be carried. NFPA Standard No. 19 recommends that 1,500 feet of 2½-inch, double-jacket, rubber-lined hose be carried on each pumper. It is best that this hose be carried in a double bed so that two lines can be dropped simultaneously. With a double bed, it is also possible to have one lead of 3-inch hose and one of 2½-inch hose. The water-carrying capacity of a 3-inch line is half again that of a 2½-inch line. The 2½- or 3-inch line is for attack on the exterior of the fire building or on the interior of a larger building where fire spread and intensity are extensive. For inside attack on a normal fire in a normal occupancy, NFPA Standard No. 19 recommends that 400 feet of 1½-inch, double-jacket, rubber-lined hose be carried. Thus 1½-inch hose can be attached to the 2½-inch hose for a direct attack on the inside fire. There are two definite advantages to the use of the 1½-inch line on an inside fire. First, there will be less possibility of water damage from an excess of water. Second, the 1½-inch line is much easier to maneuver and the fire area can be covered with greater ease and with less manpower.

Fire hose normally used by fire departments consists of a rubber tubing covered by a woven jacket of two thicknesses. For years the outer woven jacket was of cotton. Recently, this outer jacket to some degree has been replaced by a polyester fiber or a mixture of polyester fiber and cotton. There are advantages and disadvantages to both. Cotton-jacketed hose is heavy and tends to mildew when damp; therefore it must be dry before being bedded on the apparatus. The biggest advantage of polyester fiber is in its lighter weight and minimal tendency to mildew or rot. Its chief drawback is a tendency to kink in use and thus to retard the flow of water.

Fire hose that receives little use can be constructed with only one woven outer jacket or be unlined. Unlined linen hose is common for interior standpipe house protection or brigade use. This type of hose tends to leak water, and so is suitable for use where hot ashes or embers may come in contact with it—for example, in forest firefighting. Friction loss, however, is considerably higher in un-lined linen hose.

Standard sizes of rubber-lined, jacketed hose usually run from 1½ inches through 2, 2½, 3, 3½, 4, 4½, and 5 to 6 inches. So-called booster-line hose is usually ¾ inch or 1 inch in size. This is a rubber-lined, rubber-covered hard line usually carried on reels, supplied by the water tank and useful on the small exterior or interior fire.

To draft water from a static source, the fire hose or "suction" hose must be reinforced so that when the interior pressure is reduced, the hose does not collapse from the outside atmospheric pressure.

Weight is a limiting factor in the size of hose being used, 3 inches being about the maximum that can be efficiently used as a hand-held line. Larger-sized hose is restricted to use in fixed appliances or as supply lines. The following table is for rubber-lined, cotton, double-woven hose in the standard 50-foot length.

SIZE OF HOSE	1½ INCHES	2½ INCHES	3 INCHES
Weight dry	32 pounds	58 pounds	84 pounds
Weight filled with water	70.3 pounds	164.2 pounds	236.9 pounds

Figure 9-4   Modern 1,000-gpm pumpers. Both these pumpers carry a booster tank with two reels of 1-inch hard rubber hose, generator and floodlight, a three-channel radio at the engineer's panel, and enclosed compartments for tools and fittings.

Courtesy of the Chicago Fire Department.

Many departments preconnect 1½-inch hose to pumpers equipped with water tanks. As the pumper stops at the fire building, this preconnected line can be led out and water applied instantly from the tank. Before the water tank has been emptied, a preconnected line of 2½-inch or larger hose connected to the pump intake port can be hooked up to the closest hydrant supply or to a larger water tank that responds with the pumper. A supply may also be obtained from a second pumping engine in relay from an ample source of water. For backup if needed, a 2½-inch hose may be led out from the pumper in close proximity to the seat of the fire. (A 2½-inch hose for initial hand-line attack feeding master streams may also be preconnected to the pumper equipped with a water tank for instant knockdown or slowdown of larger fires.)

Those pumping engines that have water tanks usually also carry one or two reels of ¾-inch or 1-inch rubber-covered hard-line hose. This hard line can be stretched to throw instant water on the automobile, prairie, rubbish, or one-room fire.

The pump with larger pumping capacity, in the 2,000-gpm class, should provide for three beds of hose. In addition to 2½-inch and

3-inch hose, this class of pumper should carry some 3½-inch hose. This size of hose is necessary to cut down on friction loss if pump capacities are to be fully utilized. Because of the ability to pump large capacities, a good water supply is essential.

The pumping engine must also carry some footage of 4½-inch, 5-inch, or 6-inch hose to be used as a supply line to the intake port of the pumper when hooking up directly to a hydrant or a water source. If water is to be drafted, the intake line must be a hard suction to prevent collapse when the vacuum necessary for suction is created in the intake line.

Hose Stream Nozzles

The pumping engine carries a varied supply of nozzles to deliver the water in an effective stream to the fire area.

The nozzle may be a straight-stream type only, a spray- or fog-pattern tip only, or a combination of straight-stream and spray. A better solid-stream or spray pattern will be obtained from the nozzle that has a tip that serves one function only. But the advantage in having a nozzle that can be instantly changed from a solid to a spray pattern while being used makes the combination nozzle more popular than the other types.

Solid-stream tip sizes vary from ¾-inch — approximate delivery, 120 gpm at 50-pound nozzle pressure — to a 1½-inch tip — approximate delivery, 560 gpm at 70-pound nozzle pressure. The booster line normally will carry a combination nozzle that delivers approximately 30 gpm. The 1½-inch hose used in an inside attack will also normally carry a combination nozzle. The smooth-bore solid-stream nozzle is most effective in an outside attack on a large fire.

Minimum Equipment

Various basic equipment and fittings are specified by NFPA Standard No. 19 (which contains a complete listing). A partial listing includes:

1. A 14-foot roof ladder
2. A 24-foot extension ladder
3. A fire axe

4. An approved class A type of fire extinguisher — usually a 5-gallon water extinguisher, and a class B extinguisher — usually a multipurpose dry chemical.

Other equipment and fittings will be carried by the individual fire departments that need them in their basic hose lead-outs and specific response problems.

LADDER TRUCKS

The pumper is rightfully said to be the attack force of fire extinguishment; all other equipment is merely an aid to the pumper in the attainment of extinguishment. Yet so important is the work of the ladder company that most fire commanders agree with the theory that all ladder truck–assigned firefighters, or, as more commonly called, truck men, should serve first on an engine company before receiving the truck assignment. With this prior experience, they would have firsthand knowledge of the effectiveness and need of truck aid to the engine company.

This truck aid in essence helps to meet the need for effective ventilation, forcible entry, and assistance in the advancement of hose lines. Of course, the primary task of the truck company is the saving of life, a task which often depends on just such aid.

Ladders are an essential item in both lifesaving and fire control. The modern fire ladder truck is usually equipped with a 100-foot hydraulic-lift metal aerial ladder (shown in Figure 9-5). Aerial ladders have evolved from 65-, 75-, or 85-foot wooden spring-raised aerials. The older ladder truck had a two-piece aerial ladder that required a tiller wheel at the rear because of the length. The fly had to be manually raised and lowered into the fire building. A later development of the wooden aerial ladder was a hydraulic-lift device that greatly speeded up the operation of getting a ladder into proper position. Some wooden aerial ladders, both spring- and hydraulic-operated, are still in service and performing the job required.

The modern hydraulic-lift aerial ladder is in three or more sections, and the tiller wheel has been eliminated. Ground jacks are provided on each side of the apparatus. In operation at the fire scene, they absorb part of the ladder load and prevent swaying or tilting toward the building side when the ladder is raised.

A complement of ground ladders is also carried on the ladder truck. Standards recommend that the well-equipped ladder truck

Figure 9-5 A modern aerial ladder truck with four-sectional 100-foot aerial. The truck is equipped with its own water tank, booster pump, and reel of 1-inch hard rubber hose. It also carries its own generator and floodlights.

Courtesy of the Chicago Fire Department.

carry approximately 200 feet of ground ladders (see NFPA Standard No. 19), including one-piece straight-frame ladders, extension ladders, and roof ladders. The modern trend is for these ladders to be of aluminum construction. The lighter weight makes for greater ease in handling and raising, but it poses the problem of the ladder's being blown down by gusts of wind and the ever-present problem of its striking live electrical wires while being raised.

Laddering a building is only one phase of the duties of a ladder truck. Equipment must be carried to effect forcible entry, to open floors, walls, and ceilings to uncover hidden fire, and to open roofs, ceilings, and walls for ventilation purposes. Tools used in these operations are pike poles, pry bars, axes, and power saws. Other equipment carried is hand water pumps; chemical extinguishers; and 1½-inch-tip or larger master-stream appliances, such as aerial pipes or turret guns. The aerial pipe is an effective aid in combating the exterior fire in the multistoried building. (Elevating platforms are discussed later in this chapter.)

Some ladder trucks are also equipped with a water tank, a small pump, and a 200-foot reel of 1-inch hard rubber hose. This equipment is useful for the small exterior fire or for a fast attack on the large fire in a holding action until more powerful streams can be put to work. This type of apparatus has a combined use and may be known as a "combination."

In some municipalities, the ladder truck may be equipped with a pump of larger capacity and also carry its own bed of 2½-inch hose. It then becomes a quadruple fire apparatus or a "quad"— a pumper-ladder truck. A combination or a quad may or may not be equipped with an aerial ladder.

The ladder truck also may carry salvage equipment and specialized rescue equipment. They are discussed in the sections on salvage and rescue units.

SUPPLEMENTARY EQUIPMENT

Squad-Rescue Units

Several other pieces of equipment assume a role similar to that of the ladder truck — the role of entry, ventilation, opening up, and general fire duty. Their greatest difference is that these units do not carry the complement of ground ladders or the aerial ladder. They are supplementary units, easier to equip and able to perform equally well, using the ladders of the ladder truck that precedes them to the fire scene. They will usually carry extra equipment, such as cutting torches, additional self-contained air masks, possibly a power winch, and lifesaving equipment including resuscitators and inhalators. Generally, they also may bring a gasoline-powered saw, an air hammer and torch, and cutting tools. (See Figure 9-6.)

The manpower operating such a unit can vary, depending on the policy of the municipality. Three or four men can operate the equipment, but if the unit is to be considered also as a manpower pool for use on other units as needed at the fire scene, one or two more men would be more desirable.

In those departments that do not have specific squad or rescue units, the equipment and duties will be assigned to the truck company.

Figure 9-6 Squad-rescue unit. This all-purpose type of apparatus uses its own turret master-stream appliance. Its compartments store specialized equipment.

Salvage Unit

The salvaging of merchandise and equipment is an important function of the fire department. To this end, many departments equip salvage units whose primary task is to prevent water damage by the covering of stock or the diversion of water. Figure 9-7 shows a salvage unit at work. The special equipment includes salvage covers, water vacuums, squeegees, brooms, shovels, and mops. Where the need for salvage work is not pressing, the personnel on these units can be used as engine or truck men. If the factors of economics and need militate against a salvage unit, this work and equipment become part of the responsibility of the truck company.

Elevating Platforms

One of the most recent developments in fire-fighting equipment is the fire apparatus equipped with an elevating platform commonly

called a snorkel unit (illustrated in Figure 9-8). It differs from the elevating platform used for tree-trimming, material hoisting, and so on, as it is equipped with a jointed riser that can be fed by two or more hose lines at the base for the operation of a master-stream appliance at the platform basket. Elevated platforms have been designed that are capable of reaching heights of 50 to 150 feet. They have virtually replaced water towers and are more effective than aerial ladder pipes, although the latter still has a place in fire service.

The maneuverability of the elevated platform permits it to operate over the top of the fire. It is also capable of sweeping over a wide front of fire. Moreover, elevating platforms play an important role in rescue operations. They can cover some locations or portions of certain structures that are difficult to ladder or reach with aerial ladders. There probably is a feeling of greater security when de-

Figure 9-8 Elevating platform. This particular "snorkel" is equipped with its own pump unit, note direct hookup to hydrant, and carries its own reel of 1-inch hard rubber hose. The platform can be operated from the rear at ground level, or by controls in the basket.

Courtesy of the Chicago Fire Department.

scending in a basket than when clambering down ladder rungs. Also, an injured or unconscious victim can be lowered in a stretcher in the snorkel basket.

Where many victims must be evacuated from a single spot location, the aerial ladder will prove more expedient. It can be raised to a given location faster and will remain fixed until all victims have been removed. The elevating platform must be lowered when the basket's capacity limit is reached.

The elevating platform apparatus can be designed to carry a complement of ground ladders, or it can be equipped with a pumping unit and its own complement of hose.

SPECIAL UNITS

Discussion thus far has centered chiefly on equipment designed for the application of water on the fire (with some description of rescue and salvage equipment). In urbanized areas, there may be a need for special units that can apply extinguishing agents on class B and C fires — in particular, on airport fires or high industrial hazards.

Dry Chemical Unit

The dry chemical unit is designed to carry a large quantity of dry chemical, usually potassium bicarbonate (purple K), that can be expelled by the pressure of nitrogen gas through a reel of 1-inch hard rubber hose. The standard unit is 750 pounds, offering good extinguishment potential for the class B flammable liquid fire.

The danger of a possible rekindle of the fire is inherent in the use of a dry chemical as the extinguishing agent, however. A dry chemical has little cooling effect. When it is dissipated, a rekindle may occur from hot spots or general latent heat. (See Chapter 8 on extinguishing agents.)

Therefore, the dry chemical unit usually carries a water tank with foam attachment. The foam generated is expelled through a second reel of 1-inch hard rubber hose. This one-two punch of dry chemical and foam offers excellent control and extinguishment for the difficult flammable liquid fire. Another precaution is necessary in this twinned attack. The foam must be compatible with the dry chemical to retard foam breakdown (also discussed in Chapter 8).

Crash Wagon

Large-airport fire protection requires equipment with the potential of fast extinguishment for the low-level impact aircraft disaster. The vehicle must be able to carry sufficient extinguishing agent and an efficient dispensing system. It must possess speed capabilities and power enough to carry this equipment to off-runway crash sites. (Figure 9-9 shows an airport crash wagon.)

The airport foam-generating crash fire truck in service at large municipal airports carries its own water supply of 2,500 to 3,000 gallons. Two 250-gallon tanks carry 500 gallons of foam. The initial attack with the foam can be made with hand-held hard lines or vehicle-mounted nozzles. A turret-mounted nozzle is capable of delivering 500 to 1,000 gpm of foam in either a straight-stream or fog pattern. The straight-stream range is 150 feet.

Figure 9-9 An airport crash wagon. This unit is capable of traveling over rugged terrain and delivers a sustained foam attack. Newer equipment carries "light water" and dry chemical in a twinned line, dual-nozzle attack.

Figure 9-9 An airport crash wagon. This unit is capable of traveling over rugged terrain and delivers a sustained foam attack. Newer equipment carries "light water" and dry chemical in a twinned line, dual-nozzle attack.

Courtesy of the Chicago Fire Department.

Additional quantities of foam are carried that can be used with a supplementary supply of water brought by water-tank truck or hydrant lines.

The airport crash unit should have the potential of applying a minimum of five minutes of extinguishing agents. After this period, backup equipment should be arriving. This supporting equipment helps the foam-extinguishing crash truck to extinguish fires in spaces where foam cannot penetrate. The dry chemical unit or a cardox (carbon dioxide) unit also can be used at the airport fire station.

Cardox Unit

The cardox — CO_2 — crash fire unit is a low-pressure storage tank of CO_2. The carbon dioxide is maintained at a temperature of $0\,°F$ under 300 psi pressure. The liquid CO_2 can be discharged through hard lines or a boom nozzle. Discharge may be 1,250 to 2,500 pounds per minute. As with the foam crash wagon, the potential to blanket an area rapidly must be present and maintained if trapped victims are to be rescued.

Other Specialized Equipment

Urban departments may also have other units that perform a particular task in the department's fire-fighting and rescue plan of operation. A description of some of them follows.

A COMMAND CENTER UNIT

Communications are a vital concern at the fire scene. They must be maintained between units at the fire ground and also between the communication center at the fire ground and the departmental communication center at headquarters. The on-site communication center may be the fire command chief's vehicle, but a larger command center unit is of great aid at large fires.

This unit can tabulate lines led out from working pumpers, determine the ability of working pumpers to handle extra lines, and check water supplies and pressures available. The specific tasks being performed by the working units can be plotted on a plat (map) of the fire building.

Full and detailed information on the fire building should be available in the unit's files, as well as information on other specific hazards that may be encountered at any fire scene.

LIGHT WAGON AND ELECTRICAL GENERATING UNIT

The large nighttime fire can be simplified by the use of banks of floodlights mounted in a strategic location. Power may be needed for the use of power tools and other electrical equipment.

This need can be supplied by a unit carrying an electric generating plant. Banks of floodlights can be mounted on the apparatus for exterior lighting. Interior lighting can be supplied by portable floodlights carried by this unit.

SMOKE EJECTORS

A special unit with a powerful suction fan and large-diameter flexible tubing can be employed to suck out the smoke and gases from the fire building, in particular, from basement areas.

FIREBOATS

Fireboats can be employed in most localities having harbors, piers, and wharves. The unit may vary from a small jet-propelled rescue craft to a large tug. Pumping capacities can range from 500 to 14,000 or more gallons per minute. These units afford protection to vessels, wharves, and buildings along the harbor or water's edge and can also relay water inland to other units at a fire scene.

In conjunction with jet-propelled rescue crafts and fireboats, a fire service may include teams of trained scuba divers. The teams can be used in rescue and recovery from water accidents. The drowning victim, the ship incident, the automobile or plane in the water can be quickly and efficiently handled by the well-trained scuba team. Scuba fire-attack teams are also used in fire control and extinguishment on ship, pier, and wharf fires.

HELICOPTERS

Helicopters have proved themselves a valuable ally in search-and-rescue missions both in the armed services and in private life. Their role in the fire-fighting service is fast being developed.

The use of helicopters in the high-rise fire is an area of extreme concern to the fire service. Many victims who otherwise would have lost their lives have been rescued by helicopter from high-rise building fires. Also of great significance, an accident victim can be quickly transported to a hospital from an otherwise inaccessible area or without being subjected to congested street traffic. And helicopter tankers are used to drop chemicals in forest firefighting.

The fire-fighting service has yet to realize the full potential of the helicopter for quick size-up, for overview for command decisions, and for its possibilities as an extinguishant carrier. In the high-rise fire, it offers unlimited possibilities if further aids can be found to be used in conjunction with it.

ADAPTATION TO CIRCUMSTANCES

Many smaller departments cannot afford the luxury of special and specialized equipment. Yet the need for the services made possible by this equipment must be met. This need can be somewhat alleviated by the assignment of these specific tasks, such as salvage work, to other units. Such equipment as portable smoke ejectors and electrical generators can be carried by a squad unit. Minimal supplies of foam powders and liquids can be carried for use in portable foam hoppers or eductors. Portable ground master streams or aerial ladder pipes can be employed in place of the articulating boom or turret-wagon equipment. Many of these aids may be employed in combination units, some of which have been previously discussed.

Ambulances

The fire department is dedicated to the saving of life and property. Too often one group of persons in urgent need of emergency service is overlooked. This group includes the street-accident victim, vehicular and other, and the accident victim within a building. These victims are in immediate need of first aid and removal to a hospital to prevent death from loss of blood, breathing difficulties, or other physical injuries.

Too often, and especially in the case of the street-accident victim, the police must respond first. If in their opinion an ambulance is needed, the call is then sent out to a private ambulance service.

Many fire departments maintain an ambulance for emergency service. More should do so. The accident victim deserves the life protection afforded by a public ambulance that can respond immediately, often arriving on the scene before the police and providing quick removal to the nearest hospital for immediate care.

BREATHING EQUIPMENT — RESPIRATORY PROTECTION

The firefighter faces an atmosphere loaded with contaminants, both particulate matter and gases.

Masks

NFPA Standard No. 19 recommends that each pumper unit carry two self-contained breathing masks and that each hook-and-ladder truck carry six self-contained breathing masks. The masks must be approved by the U.S. Bureau of Mines.

In April 1971, the NFPA removed all service-type N masks from its list of approved equipment. This canister type of filter mask was widely used in the fire service after World War II. It was light in weight, cheap, and had a fairly long life. It filtered out smoke and smoke-particulate matter and small concentrations of certain gases. But it did not completely filter out carbon monoxide, a big hazard to the firefighter. It had to be used in an area having sufficient oxygen, as it did not generate or store any oxygen. Too often, the fire scene has an oxygen deficiency that can affect the firefighter's judgment or cause unconsciousness. The caisson or tunnel may also have an oxygen shortage, and firefighters attempting rescue here can be trapped themselves because of the oxygen deficiency.

New products of today's technological advances also may offer hazards that cannot be filtered out by the filter type of mask.

For these reasons, NFPA Standard No. 19 recommends that only self-contained breathing equipment be used in the fire service. Self-contained breathing apparatus is of two general types.

REPLACEABLE CANISTER MASK

This type of mask uses a canister that removes the exhaled carbon dioxide and moisture and generates more oxygen. It can be used in oxygen-deficient areas and in atmospheres of high toxicity. It has a canister life rated as one hour.

This type of mask is activated only after a seal on the canister is punctured. In a recycling process, the exhaled breath passes through this canister and a breathing bag. Some exhaled oxygen is salvaged and returned to the user. More oxygen is generated through the reaction of chemicals in the canister and the exhaled breath.

CYLINDER OR "DEMAND" MASK

In this type of mask, the facepiece is connected to an air or oxygen cylinder tank under pressure. A regulator or reducing valve regulates the pressure and flow to the user. (Figure 9-10 shows such a mask.)

Several years ago, the oxygen cylinder mask seemed to be the most popular. This usage was probably tied in with the availability of oxygen to refill cylinders and the lack of availability of pure air. The sudden growth of scuba-diving equipment has made pure air available, and brought a sudden switch to the air cylinder, however. It is now used almost exclusively. Also, the fear of oxygen as a contributor to combustion in a mishap lends more confidence to the use of the air cylinder mask.

The normal tank has a 30-minute supply for the normal user. However, under excitement, stress, or heavy labor, actual use time

Figure 9-10 Breathing apparatus. In this demand type of air mask, several sizes of cylinders may be used in the stainless steel cylinder holder.

Courtesy of Mine Safety Appliances Company.

may be reduced considerably. The rating is an average established in Schedule 13 of the Bureau of Mines.

The air cylinder demand mask is heavier and bulkier than the replaceable oxygen-generating canister type, but seems to be preferred by the fire service. The tank is carried on the wearer's back, whereas the canister is worn in front. The back position appears to be more comfortable. Further, having a ready tank of air or oxygen

seems more reassuring than merely having the materials to generate the oxygen.

The use of masks has many benefits but also some dangers. Certain precautions in their use are essential:

1.   Users must work in teams.

2.   A lifeline to the means of exit should be provided.

3.   Excessive heat and gases or liquids that are harmful to the skin must be avoided.

4.   Time must be allotted for exiting when a warning bell indicates that the tank or canister is nearing depletion.

5.   The mask user must be adequately trained under simulated conditions in the use of the equipment. Confinement in a mask is a traumatic experience in itself without the added dangers of smoke, gases, oxygen depletion, and the exertion of work.

Masks must be considered as another tool with limited uses for the firefighter (Figure 9-11). Used properly, they can not only protect firefighters' health and well-being, but also aid them in achieving earlier extinguishment or faster life rescue. Used irresponsibly, they can place firefighters in a fire situation of extreme heat, burn possibilities, or structural collapse and entrapment.

PROTECTIVE CLOTHING

The turnout coat worn by the firefighter affords some protection against heat and sparks or embers, and it sheds water. It is functional to a degree, but does not fully protect its wearer under many fire situations, and it offers virtually no protection to the firefighter trapped in extreme heat or flame.

In recognition of NASA's research and development in protective clothing for the astronauts, the association was asked to research and develop a safer turnout gear for the firefighter. Selected materials have been tested, evaluated, and fabricated into prototype designs.

The program, as of this writing, is in the demonstration stage. Fifteen city fire departments are testing both turnout and fire proximity suits. The clothing is being used for one year. If field reports justify such a step, specifications for design and fabrication will then be let out to private industry.

NASA also has developed a prototype of a new breathing apparatus with improved leakproof face mask, new harness, and improved bottle having a full 30-minute capacity.

SUMMARY

The pumper fire engine is the basic attack unit in the fight against uncontrolled fire. It is capable of delivering 500 to 2,000 gallons of water per minute. This gallonage can be delivered through a wide selection of hose sizes, nozzle diameters, and spray patterns in amounts varying from 30 gallons per minute to 1,000 gallons per minute.

However, this is only one unit in an effective attack team. The saving of life—rescue—must always be uppermost in the firefighter's mind. Where there is no life hazard, entry to the seat of the fire must be made and ventilation of smoke and gases effected.

Salvage efforts must be carried out simultaneously with fire-extinguishing efforts. That is, stock and furnishings endangered by fire and water must be salvaged, and smoke and water must be removed when the fire is out.

These necessary tasks are carried out by a ladder truck, rescue or squad unit, and salvage units that support and complement the pumper unit.

A variety of tools and equipment is used in this work, mainly ladders, forcible entry tools, and salvage equipment.

Specific units are needed for a specific task or area. For example, elevating platforms, airport crash trucks, elevating chemical units, fireboats, and helicopters serve specific needs.

Specific equipment is useful in certain instances. Respiratory equipment, lighting units, and smoke ejectors all are further tools to aid the fire-fighting force.

DISCUSSION TOPICS

1. What is meant by a positive-displacement pump? Discuss this type of pump.

2. Balance the desirable features against the handicaps of various types of centrifugal pumps.

3. What are some common sizes, types, and uses of fire hose?

4. What equipment may be found on an aerial ladder truck?

5. Name several specialized types of equipment and their uses.

6. Self-contained breathing equipment is an important adjunct to the fire service, but must be used with caution. Discuss both the pros and cons of this equipment.

RESEARCH PROJECTS

1. List all equipment carried on a pumper unit by a local fire department. Specify equipment recommended by NFPA Standard No. 19 and equipment added for local needs.

2. List all equipment carried on a ladder truck unit by a local fire department. Specify equipment recommended by NFPA Standard No. 19 and equipment added for local needs.

3. At the present time, helicopters play a supporting role in fire-fighting tactics and rescue work in certain areas. Evaluate that role and delineate a possible expanded role helicopters might play in fire-fighting and rescue work.

4. Evaluate the role of respiratory equipment in the fire service. Cite examples both for and against the use of this equipment.

5. Aerial ladders and elevating platforms are complementary adjuncts to the fire-fighting force. Both may play a similar role. List these similar task roles and the pros and cons of their use.

FURTHER READINGS

Tryon, H. George (ed.), *Fire Protection Handbook,* 13th ed., National Fire Protection Association, Boston, 1969, sec.10, chap. 3.

National Fire Protection Association, Boston:

NFPA Standard No. 19, Specifications for Motor Fire Apparatus.

NFPA Standard No. 193, Fire Department Ladders — Ground and Aerial.

NFPA Standard No. 403, Suggestions for Aircraft Rescue and Firefighting Services at Airports and Heliports.

NFPA Standard No. 414, Standard for Aircraft Rescue and Firefighting Vehicles.

Stated briefly, the role of the firefighter is a threefold one. He or she is dedicated to:

1. The saving of life
2. Protection of property
3. Extinguishment of fire

The firefighter's task in fulfilling this role has been subdivided into (1) fire prevention, and (2) fire control and suppression. The greatest expenditure of funds and manpower goes to support efforts of fire control and suppression.

But this statement may be somewhat misleading. The firefighter can be found at any local catastrophe where life is endangered, whether fire is present or not. A tornado, hurricane, or flood; an auto, airplane, or train wreck; a mine cave-in, a caisson tank disaster; a building collapse, scaffold failure, faulty elevator; any of these and more will find the firefighter responding. One such response is shown in Figure 10-1.

On a fire response he sees destruction as an ongoing phenomenon. In the nonfire emergency response, all destruction for all practical purposes has taken place before his arrival. Life safety, of course, is the first consideration; protection of property follows.

At those fires where no life is endangered, his primary concern is to confine the fire — to keep it from spreading. In his own vernacular, he will protect his exposures. The protection of exposures may be the protection of exposed property adjacent to the fire building or other floors and rooms of the fire building. He is always faced with the possibility that the fire has spread to higher floors or to hidden

FIRE-FIGHTING PROCEDURES

Figure 10-1 A firefighter at a special-duty response involving an automobile and a trapped victim. The ability to remain calm and also to improvise appropriate usage of emergency equipment is needed at most special-duty responses.

Courtesy of the Chicago Fire Department.

attic spaces. At the least, he will be faced with the prospect of smoke and hot gases that have risen above him and are in the process of mushrooming. It is only after consideration and covering of these possibilities that he can direct his attack on the main body of fire.

Yet even before these considerations, he will have carried out several important steps. And, after extinguishing the fire he will have several more steps to go before he picks up his equipment and returns to quarters to await the next call.

PROCEDURAL OUTLINE

One fire has much similarity to another fire. Fire spread repeats a pattern. This also can be said of the carpentry labor on a home or the work of a plumber installing a new bathroom. Yet, for all three tasks, each separate job is different. Upon reaching the scene and seeing the specific job conditions, the plumber and the carpenter can take fifteen minutes, a half-hour, an hour if necessary, planning the job to suit the job conditions. This time will be more than made up when actual work starts after the preplanning and layout period.

This is not true of the fire service. The firefighter has already preplanned his attack on various buildings and occupancies, but when he arrives at the fire scene and is confronted by "job" conditions, he has no time for conferences and the weighing of various possibilities. He must do something — *right now!* This need for immediate action lays stress on the importance of the fire service's following a set procedure outline that will weigh existing and developing conditions. The decisions then reached may amend the plan or assign priorities to the fire-fighting tactics and strategy to be employed.

Consideration of existing job conditions need not wait until arrival at the fire scene. They can be considered immediately upon receipt of the alarm and even before the response. The responding fire officer who will be in command at the fire scene follows a procedure pattern that encompasses eight steps, five in addition to the three steps of saving of life, protection of property, and fire extinguishment so basic to his entire intent and purpose. The procedure outline designates the following steps:

1. Size-up of the situation
2. Calling of help
3. Saving of life (rescue)
4. Covering of exposures (confinement of fire)
5. Ventilation and forcible entry
6. Extinguishment
7. Overhauling
8. Salvage

This outline does not imply that firefighting is a step-by-step procedure, nor should the inference be made that any one operation must wait upon any other. In the majority of fires, there is no life hazard nor is there an exterior exposure problem. At most fires, procedures will be inaugurated concurrently rather than sequentially. Extinguishment, if not concurrent with ventilation, may be begun before the ventilating team can "open up." Salvage work should be started as quickly as possible to mitigate possible added water damage.

The procedural outline does assign priorities to that small percentage of fires that is responsible for the greatest percentage of loss (95 percent of all fire loss occurs in 30 percent of all fires). It does aid the command officer in arriving at a plan of tactics and strategy for the peculiar conditions of this particular fire call without undue loss of time at the fire scene.

Let us explore each stage of the procedural outline.

SIZE-UP

An evaluation of the fire extinguishment procedures to be used at a particular fire must be completed by the time the apparatus pulls to a stop at the location. Size-up is an ongoing process that will continue until the last company picks up to return to quarters, but it is imperative that some plan of action be ready when the fire scene is reached. As indicated previously, this evaluation or initial size-up is a two-step procedure. Initial planning will take place upon pushout following receipt of the alarm. This planning will take into consideration factors already known and apt to affect the subsequent fire-fighting tactics; it thus will aid in the first decision made in the second step at the time of arrival. The factors to be considered at the time of response are:

1. Time of day

2. Weather conditions

3. Previous knowledge of the building or the general area

The time of day will help determine occupancy hazards. For example, life hazard in a nighttime response to a school or mercantile establishment should be minimal. A night response to a residence, apartment building, or hotel may find occupants asleep. The time of day will also affect traffic on the street and the route to be taken. Many times, the shortest route may not be the quickest when it is also the popular route of workers going to and from their jobs at the start and finish of the normal workday.

Certainly weather conditions will play an important role. On a subzero wintry day or night, there is always the possibility of a frozen hydrant or frozen water supply if the supply is a pond or river. Extremely cold weather also will affect the firefighter, who is subject to frostbite of cheeks, ears, or fingers. Ice and snow not only will delay the response, but will have an adverse effect on operations at the fire ground.

Hot weather may also be a detriment. The greater use of water during the hot spell may have greatly reduced the amount of water available for firefighting. All areas of the country cite cases of residential faucets that deliver only a trickle of water during hot weather, and of ponds and streams that tend to run dry. Again, the firefighter is not at his peak of efficiency when forced to perform his duties in the necessary protective gear under high summer temperatures. Both extremes of temperature will weigh heavily in the decisions to follow.

Wind conditions that may help to carry the fire to adjoining property or a long dry spell that favors the rapid spread of fire must be considered.

Knowledge of the building or general area is a prime aid in the formation of a plan of attack prior to arrival. If the building has been visited on a company preplanning inspection, much valuable information is at hand to be weighed while en route. The type of building construction, the occupancy, the special hazards of the occupancy, and the exterior exposure to adjoining property all will be known.

How good is the water supply? Is the fire building hard to get to? Is it in a valley or a hilly section? Are access roads narrow, steep, or unimproved? Are they blocked by train crossings? Is the building so situated on the property that it can be reached only by long and difficult hose lead-outs? These questions and many more will have been answered by the preplanning inspection and the known topography of the area.

Information received from the initial fire call may help in formulating a potential plan of attack and in determining the response. If the caller has reported an explosion, for example, an ambulance may also be started to the scene. On occasion, several calls reporting the same fire will indicate its possible magnitude. It is important that as much pertinent information as possible be obtained from the person reporting the fire.

The second phase of the size-up will be made upon arrival at, or approach to, the fire address. (See Figure 10-2.) A quick appraisal will determine:

1. Life hazard and occupancy

2. Location and extent of the fire

3. Dimensions and construction of the building

4. Possibilities of communication or spread of the fire

5. Auxiliary appliances available

6. Available water supply

Life-hazard potential can be quickly determined. At 2 A.M., the multistoried apartment building with smoke oozing from several floors or windows is indicative of a grave life-hazard situation. The single-family residence at this time of night can also pose a threat to

Figure 10-2 Consideration of factors involved in the size-up. The fire command officer, upon arrival, must quickly determine the fire ground conditions and the possible spread of fire. He then must evaluate these factors against the resources he has at hand.

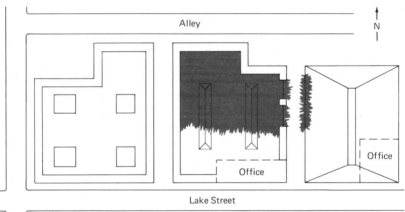

Time: 3:30 a.m.

Wind: West at 15 mph

Temp: 25°

Fire building: One-story brick, flat roof. Manufacturer of roller skates

Exposure to east:
One-story brick, truss roof. Cartage company, truck storage

Exposure to west:
Two-story brick, flat roof. Manufacturer of kitchen tables and dinette sets.

the sleeping residents who may be trapped inside. Trapped victims need not be seen at the window sill, heads enveloped in smoke, for there to be a life hazard. It is the unseen victim who must be accounted for. In most cases, a quick glance at the building, coupled with the time of day, will indicate the occupancy and life-hazard potential: a residence; a particular type of occupancy such as a supermarket; a factory, with more precise occupancy to be determined later.

The progress of the fire and direction of travel will be looked for. In the large fire, billowing clouds of smoke, open flame, or gesturing bystanders will affect the size-up before arrival at the fire scene. In many cases, the extent of fire cannot be determined until entrance is made into the fire building. Here, the dimensions and construction of the building will have been noted before the extent of fire has been determined.

Next to life hazard in importance is the extent of the fire and its possible communication to other sections or floors of the fire building, or to adjoining properties. Wind direction and velocity and building construction of both fire building and adjoining properties must be considered. Distance separating the fire building from possible exposures must also be taken into account.

A further quick appraisal will determine what auxiliary appliances are available as aids in the extinguishment of the fire. Automatic sprinklers may have either extinguished the fire or held it in check. In any event, the water supply to the sprinkler system must be augmented by the fire department pumper. Standpipes, either wet or dry, may be an aid in getting lines to the seat of the fire. These, too, must be supplied by the fire department pumper. Many industrial plants may have their own fire brigade, hose carts, or various extinguishers.

The final consideration in the initial size-up will be the water supply available. The area may be serviced with ample hydrants on large-sized mains. Or the water supply may be drawn from reservoirs, ponds, a river or lake. Water may have to be relayed from pumper to pumper or brought in by water-tank truck—an important consideration in the determination of the call for help.

CALL FOR HELP

The decision on whether to call for help depends upon information gained in the size-up of the situation. If conditions at the fire indicate that more help will be required than is responding to the alarm transmitted, the command officer must issue a call for additional help at once. In the larger cities, such a call will necessitate only the calling for a box-alarm response or for multiple-alarm response. The large city usually can adequately handle any fire that occurs. In the small town, a call for help may necessitate the utilization of all fire equipment and the recall of all off-duty men. In some areas, a mutual aid plan may have to be put into effect, with help received from other fire departments in the mutual aid network.

Quite often, the command officer in a large city will issue a call for help as a purely precautionary measure. In a doubtful situation, he knows the equipment is readily available and will call for the extra alarm. Upon the arrival of this extra equipment there is usually either a clear-cut need for the equipment and it is put to work, or the fire is under control or extinguished, and the extra equipment is returned to its quarters.

The command officer at the fire in the suburban area is faced with the problem of minimum equipment response and possible man-power shortage. The officer (even today, usually a man) may be reluctant to initiate a mutual aid call as a precautionary measure. When he does issue such a call, response time for the called companies may be long because of the distance they must travel.

Additional fire department help needed may consist of pumping engines, truck equipment, or specialized equipment such as elevated platforms, high-caliber streams, or supervisory personnel. The call for help may be for an ambulance, breathing apparatus, a smoke ejector, air compressors, or special forcible entry equipment.

The call for help may also go out to public utilities, such as the gas company, telephone company, or electric company. Where these utilities are involved in fire, it is often necessary to have lines servicing the fire building disconnected or shut off.

The help of other city agencies may be called for: the police department for traffic and crowd control, the water department for increased pressure, the sewer and street department for water removal. Where there are numerous injuries, all available ambulances may have to be called and the hospitals of the area alerted to the emergency.

To summarize: In the fire catastrophe, all available services that can be of help should be called upon commensurate with the need.

The good fire command officer will have already apprised himself of what services are available to him and the means of contacting them when the emergency arises.

Radio communication between units is an aid in directing equipment to report to a particular location or person upon arrival at the fire scene.

A minimum of time is consumed at the fire scene in the size-up and call for help. In reality, the first noticeable efforts by all members at the fire scene will be the primary responsibility of saving life.

THE SAVING OF LIFE

A preplanned inspection of the fire building will be of great aid in effecting rescues — hazards of the occupancy will have been noted and can be circumvented. Normal means of entrance and exitways will be known and supplementary means not readily apparent will have been planned for.

Peculiar to the specific situation will be the equipment used in rescue work. At the normal fire scene, it will consist of ladders, possibly a life net, masks, ventilation to aid entry, and quite possibly a charged line for backup and for maintenance of a rescue path. Life nets are rarely used in rescue work, and of even greater rarity are pompier ladders. Yet their use must be understood and considered where the aerial ladder and ground ladders cannot be used.[1] The importance of drill sessions in the use of ground ladders and the aerial ladder is readily appreciated by the firefighter who must reach occupants awaiting rescue at upper-story windows. Speed of execution is mandatory — as well as the ability to calm the panic-stricken occupant at the window. Many persons have jumped from high floors when rescue was only seconds — or inches — away. In most cases, once the ladder is in place the occupant is capable of descent but may require some assistance and moral support.

The inside rescue is much more difficult. Firefighters attempting to rescue the trapped person may come close to being overcome themselves. The victim is most often unconscious and must be physically removed by the firefighter(s). Before the firefighters can gain entrance to the building, they may have to put on self-

---

[1]The pompier ladder, or scaling ladder, is a short, single-beam ladder. The ladder rungs project both ways from the central support beam. At the top, at right angles, is a metal support that is used to "hang" the ladder from a window ledge. This ladder is designed to be used where the longest ladder available is still short of reaching the desired height. It can be raised from window to window until the necessary window is reached.

contained breathing equipment and initiate some ventilation to release some heat, gases, and smoke.

Panic engendered by the emergency is a particular problem to the firefighter. A frightened child often will crawl under a bed or table or hide in a closet. The importance of making a *thorough* search of the fire premises for victims or possible victims cannot be overstressed. The role panic has played in the large loss of life is readily attested to by the trampled bodies blocking exitways at many fires where victims have been gathered in a public place.

The role of the *engine* company in many rescue attempts is to get water on the fire building. Its members must knock down the fire and heat endangering rescue attempts and maintain a path of rescue free from heat and flame rather than attempt rescue themselves, the prime responsibility of the *truck* company.

Other life-endangering situations will require the specialized equipment carried on the apparatus asked for in the call for help. Some of this equipment may be gas and electrical shutoffs, power saws and jacks, ropes, and resuscitator equipment.

As previously stated, many of the calls to the fire department where life is at stake do not involve fire. They may relate to the natural disasters of tornado and flood, or to the accident involving car, plane, or train. The victim may be trapped in any number of ways or overcome by gas. In these emergencies, the firefighter will call upon his training and specialized small and large power equipment, breathing apparatus, ropes, and resuscitators.

COVERING OF EXPOSURES AND CONFINEMENT

After considering the life hazard, the fire command officer will concern himself with the protection of property. Where the fire has gained substantial headway, it will be necessary for him first to set up his attack as a defensive move to guard against the spread of fire to adjoining exposures. Thus he will confine the fire to its present area wherever possible. His first consideration will always be to protect the property not yet involved in fire.

However, in many cases where the fire is relatively confined or is just reaching exposures, he will decide to follow the criterion, "The best defense is a good offense," and launch a direct attack upon the seat of the fire. By this direct attack he will also protect the exposures. This is not to minimize the importance of exposure coverage.

Where the fire is extending and possibly making its way through a structure by means of concealed chases, the covering of exposures must get first consideration. Little will be accomplished if the fire is extinguished at its point of origin and later is found to have traveled through concealed passageways to upper parts of a building, involving larger areas of the building and making total extinguishment that much harder.

As previously cited, there are two types of exposures to be covered, interior exposures and exterior exposures. (See Figure 10-3.) Interior exposures constitute the hazard that occurs within the fire building itself. This interior-exposure problem may be the vertical exposure to fire involving communication of fire from one floor to another by an open stairway, pipe chases, or other floor-to-floor openings. The exposure hazard may be a horizontal problem of fire

Figure 10-3   Problems of exposure and the communication of fire. Here, the firefighter is faced not only with the communication of fire within the fire building but also with the fire's spreading to an exposed building—with possible life hazard.

spread from one room to another or from one end of a building to the other on the same floor level.

Exterior exposures involve the spread of fire from one structure to an adjoining one. Fire may communicate by flying embers from a great distance, by radiated heat, or by flame communication through roof, wall, or window of the original fire building.

Covering of exterior exposures is usually a large task in that the fire must have gained considerable proportions before it can endanger surrounding buildings. This operation almost invariably necessitates the use of heavy streams on exposures or on the fire building and may require the services of a good many companies.

In the event of flying embers, it may be necessary to patrol an area and use booster lines to control hot embers falling on adjacent properties.

Interior-exposure coverage necessitates hard work and common sense. It is mandatory that the floor above the fire be explored so that possible communication of fire can be checked. Plumbing walls, in particular, and ceilings may have to be opened up to ensure that the fire has not communicated. Fire communication takes peculiar paths and a basic knowledge of building construction is needed. A fire wall is no guarantee of complete fire stoppage. A carpenter may have knocked down a brick and let a joist protrude through to provide a fire path. A plumber or an electrician may have knocked a hole for passage of necessary pipes and then failed to seal the opening. A fire stop may have been omitted between joists or wall studs. These examples should illustrate the importance of exposure protection.

Interior-exposure protection necessitates good ventilation practices and possibly forcible entry.

VENTILATION AND FORCIBLE ENTRY

The saving of life, prevention of the extension of fire, and facilitation of reaching the seat of the fire for rapid extinguishment are quite often predicated on the observance of good ventilating procedures. While the fire command officer is ordering his lines into position for exposure protection, confinement, or extinguishment, he will also be ordering his ventilation and forcible-entry team into action. (Figure 10-4 shows a firefighter working on ventilation.) He

Figure 10-4 A truck man ventilating a fire building. He is using a power saw to open the roof above the fire to ventilate smoke, heat, and gases.

Courtesy of the Chicago Fire Department.

must observe one caution: Due to the possibility of a flashover when a greater supply of oxygen is introduced into the fire building, he will not initiate forcible entry or ventilation until a charged line is on hand. The one exception may be in the extreme case where he is attempting to draw hot gases and smoke away from trapped victims.

Where it is necessary to use forcible entry at a building, all that may be necessary is the use of a pinch bar, axe, or pry bar on an exterior door. Other cases may require electric, gasoline, or hydraulic power tools, torches, burning bars, or possibly even explosives.

The problem of forcible entry, as well as ventilation, has increased in those remodeled buildings where windows and store fronts have been bricked up. Many stores have iron grillwork or roll-down steel shutters that block entry during nonbusiness hours. They may require the use of winches to force entry.

Ventilating procedures are usually directed toward:

1. Possibly saving life (or lives) by drawing heat, smoke, and fire from trapped occupants

2. Preventing the spread of fire through mushrooming or vertical or horizontal spread

3. Aiding the engine company to reach the seat of the fire for extinguishment

Having forced entry and being able to hold that entry because of good ventilating procedures, the command officer can now get on with his original intent — the extinguishment of the fire.

EXTINGUISHMENT

Extinguishment and extinguishment procedures, like other fire-fighting steps, will vary according to the type of fire and the material burning. Attack upon the structural fire will be based upon the extent of fire, height, area, construction, and occupancy involved. The fire may be of the exterior type, such as in a lumberyard or grandstand. Other possibilities are waterfront fires of ships or docks; and aircraft fires, either on or off the field, in congested or uncongested areas.

Extinguishing agents may vary from a 5-gallon water pump to heavy-duty large-caliber streams. Dry powder, dry chemical, foam, or CO_2 might be the extinguishant.

Amount of equipment can range from a single pumper response to a call for assistance in a mutual aid pact. Extinguishment will be covered more fully in Chapter 11 on fire-fighting tactics and strategy.

Thought must be given to the proper positioning of all necessary equipment as it arrives and reports—and the matter of reporting is an important one. On occasion, the extra response company sees a position that needs covering and goes to work without reporting. However, the command officer who has been on the scene and has called for the additional equipment may have another plan of attack in mind—a plan that is sidetracked by the misplacement of a piece of equipment he was counting on.

With sufficient equipment and proper placement, extinguishment will only be a question of time, dependent on the magnitude of the fire.

However, before total extinguishment is effected, a certain amount of overhauling of the building and its contents will be necessary.

OVERHAULING

When the fire command officer has determined that the fire is under control and that only the mopping-up operations remain to total extinguishment, he will give thought to releasing some equipment at the scene and returning the units to service in preparation for another alarm of fire at another location.

When most of the firefighting has been from the exterior, lines will have to be advanced to the interior for final operation. Great care and judgment will have to be exercised and an evaluation made of the structural soundness of the building. In some cases, the structure may be declared unsafe and no one should be permitted to enter. However, in most cases, lines will be advanced to the interior for structural and stock overhauling. Ceilings, walls, floors, any possible hidden space, will have to be opened up to expose undetected fire and to prevent possible further communication. This opening up may have taken place as a part of, or will now become a part of, covering interior exposures.

Stock will have to be moved to uncover burning embers. Salvageable material will be moved to a safe location. Unsalvageable material will be checked and in some cases removed from the building. Caution must be exercised so that no stock or material deemed salvageable or meriting the owner's checking for valuables is thrown out. Utilities must be checked for damage, and where fire damage has occurred to the system, water, gas, and electric lines will be shut off where they enter the building.

As a part of the overhaul procedures, a cause of fire origin will be sought. Any indications of arson must be preserved for further scrutiny by arson-investigating units.

Water used in the extinguishment of the small amount of fire remaining will be supplied by small-stream appliances to minimize further water damage to the fire floor or floors below. In overhauling deep-seated fires or closely packed materials such as cotton bales or rolls of paper, wet water additives can be used. The greater penetration obtained by their use will cut down the amount of water needed and shorten the overhaul time. This overhauling, as well as some previous procedures taken, is a part of salvage.

SALVAGE

The complete operation of the fire department at the alarm of fire has been with the intent of salvage in mind — to salvage the life of any endangered person and to salvage much of the building and its contents from the ravages of the fire. (See Figure 10-5.) This mental attitude is inbred or instilled in every firefighter. It will automatically direct his operation to:

1. The quickest extinguishment

2. The least expensive way of forcing entry

3. The least expensive way of ventilation or opening up to expose fire

4. A minimum use of water

5. All due care of contents and furnishings by
a. Covering furniture, stock, and equipment with salvage covers
b. Making catchalls for water seeping through on floors below the fire floor or directing this water harmlessly off the premises

As overhauling and final extinguishment are being accomplished, the firefighter will drain off excessive water on floors and possibly syphon out water from below-grade levels. As his final step in the overhauling process, he will salvage whatever has value — sentimental or monetary. He will complete ventilation with forced blowers where necessary to remove remaining smoke and some mois-

Figure 10-5 A salvage squad covering stock not directly involved in fire, thus protecting it from smoke and water damage. This salvage work on floors not directly involved in fire must be undertaken simultaneously with extinguishment efforts on the fire floor.

Courtesy of the Chicago Fire Department.

ture from the fire area. Automatic sprinkler systems that have been activated will be restored to service.

As the last length of hose is being picked up, he will have closed up the small openings in the roof, either fire-burned or opened for ventilation purposes, with tarpaulins or tarpaper. Windows will be cleared of broken glass remnants and covered with plywood or tarpaper where feasible. The building will be returned to its rightful owner and occupants in the best possible condition under the circumstances.

As the last company returns to its own quarters, its members will share the feeling of having protected life and property to the best of their ability, using equipment available and standard procedures and tactics.

SUMMARY

The fire department has a threefold objective:

1. Saving of life
2. Protection of property
3. Extinguishment of fire

To obtain these objectives necessitates a routine procedural pattern that will ensure that all necessary steps be taken and that these steps become second nature.

Eight-Step Procedure

1. *Size-up of situation.* Initial planning; formulation of plan of attack; determination of the need for additional help.
2. *Calling for help.* The calling for additional fire equipment and special units, air compressors, smoke ejector, and so on. The call may be to gas and electric utilities, the police, the water or sewer department, or other municipal agencies.
3. *Saving of life* (the primary goal of the fire department). At the fire scene, normally the use of ladders or, in rare instances, a life net; masks for interior rescue, aided by proper ventilation and a charged line to knock down heat and fire. In other situations involving life (vehicular accidents, gas or electrical involvement, unconscious victim in tunnel or tank), the use of specialized equipment — power tools, ropes, resuscitator equipment.
4. *Covering of exposures and confinement of fire to the present fire area.* In a fast-spreading fire, a defensive move (the covering of exterior exposures) to protect adjoining properties. A fire involving one building still needs covering of all the interior areas to keep fire from spreading within the fire building.

In particular, the fire must be prevented from traveling to upper floors. This requires the covering of interior exposures. In many cases, the best covering of exposures is a direct attack on the fire seat itself.

5. *Ventilation and forcible entry.* Good ventilation procedure, consisting of creation of vents in a burning structure to remove noxious fumes and gases so as to expedite rescue operations, to facilitate fire control and extinguishment, and to prevent or

minimize the spread of fire. Possibly, forced entry into the fire building with the use of equipment at hand. Effective performance of this step often determines success or failure in achieving the department's threefold objective, given earlier.

6. *Extinguishment techniques varying with different fires.* Extinguishment to accompany or follow covering of exposures and to be completed once entry is available and heat, smoke, and gases no longer bar the way.

7. *Overhauling.* Removal of stock and furnishings and opening up of walls and ceilings to detect hidden embers and small pockets of fire remaining after general extinguishment.

8. *Salvage.* Protection of stock and furnishings from fire and water damage with covers while the fire is in progress. After fire is under control, removal of salvageable stock and furnishings to a safe area, removal of excess water from building, and stacking of débris piled in a central location.

DISCUSSION TOPICS

1. Why is the size-up basically a two-step procedure?

2. What agencies may be called for help at the fire scene—other than fire department equipment?

3. What is entailed in the phrase "cover your exposures"?

4. When can it be truly said that the fire is out?

5. Overhauling and salvage have much in common. Discuss the similarities and differences.

RESEARCH PROJECTS

1. List supplementary equipment of both the fire department and other agencies in the local area that can be used by the fire department in an emergency situation.

2. List the public utilities in your area. Include their telephone numbers, and the equipment and services available from them in an emergency.

3. Privately owned businesses use much equipment that can be used in an emergency. List some equipment owned and operated by private business that can be leased or hired if needed by the fire department.

4. Assuming a catastrophe involving many injured and dead, list local agencies and facilities that can be called upon in the emergency. Include other municipal bureaus and functions.

5. Write in depth on one of the basic procedures outlined and discussed in this chapter.

FURTHER READINGS

Kimball, Warren Y., *Fire Attack—1,* National Fire Protection Association, Boston, 1966.

Kimball, Warren Y., *Fire Attack—2,* National Fire Protection Association, Boston, 1968.

Layman, Lloyd, *Fire Fighting Tactics,* National Fire Protection Association, Boston, 1953.

Walsh, Charles V., *Fire Fighting Strategy and Leadership,* McGraw-Hill, New York, 1963.

The art of firefighting is analogous to the art of military science. Many terms and the hierarchy of command are borrowed from the military. Yet there is a difference. The military can press its attack against the foe at a time and on a battlefield of its own choosing. The fire service must seek its objectives against a foe that names the time, place, and conditions of the encounter. The planned strategy and tactics of the fire service thus can often be negated by weather conditions, inaccessible areas, and extent of fire.

Strategy may be defined as the science of forming and carrying out operations — planning and directing. Tactics is defined as the science and art of disposing forces and performing evolutions when needed.

Strategy in the fire service has become the formulation of, and practice in, drill evolutions and the preplanning for a possible emergency. Tactics becomes the fulfillment of the strategical plan under emergency conditions by the individual unit under supervision of a command officer. Strategy is the bag of tools — tactics, the choosing of the proper tool for the job.

The importance of a good training program cannot be sufficiently emphasized. There is no time for second thoughts or casual planning on the fire ground — a plan of action must be immediately put into being. These tactical maneuvers must have been formulated into training evolutions, and their execution must have become second nature through a good training and drill program. In many instances, the strategical plan may need changing as the tactical situation changes, but with a good base of drill evolutions and the ability to execute these evolutions, its adaptation becomes a simple

FIRE-FIGHTING TACTICS & STRATEGY

matter. The value of the good training program may be summed up in the one word "time" — the *time* necessary to get water on the fire.

The standard time-temperature curve gives an indication of the high temperatures that can be reached in a matter of minutes — 1000° in five minutes, 1300° in ten minutes. Where temperature has reached the point when flashover can occur, much damage will have taken place before the fire department can gain control. Where the area is large enough to require multiple streams before control and extinguishment can be attained, the building and its contents for all practical purposes may have to be written off as a total loss.

When the time en route and the time needed to get water on the fire are kept to a minimum, or when temperature buildup is slower than the standard time-temperature chart, the fire department will have an excellent chance to stop the fire with only minimal loss.

Most fires can be controlled with a 30- to 100-gallons-per-minute flow. A good working fire requires 400 to 1,000 gallons per minute. Any fire requiring flows beyond this rate can be considered a large fire with probably large loss.

The 1973 grading schedule of the Insurance Services Office requires that fire flows be determined at appropriate locations throughout the municipality. (Fire flow is defined as the rate of water flow needed to confine a large fire to a block or complex of buildings.) The minimum fire flow requirement is 500 gallons per minute and the maximum for a single fire is 12,000 gallons per minute. (The stockyard fire in Chicago in 1934 had a fire flow of 50,000 gallons per minute.)

Required fire flows take into consideration structural conditions and building density. In general, the following guidelines can be used in estimating required fire flows and the number of lines needed for control in residential occupancies.

TYPE OF CONSTRUCTION	REQUIRED FLOW
Stable residential neighborhood, single family	500 gpm
Large residential or multiple-story	1,000 gpm
High-value residences and apartments	1,500–3,000 gpm
Densely populated, three-story or higher	up to 6,000 gpm

Table 11-1   Standard Flows in Gallons per Minute
for Smooth-Bore Nozzles

With nozzle pressure of more than 50 pounds and a nozzle diameter of 1⅛ inches or more, the stream must be considered a master stream and cannot be hand-held with safety. Adjustable solid-stream spray nozzles will usually be rated in gpm by the manufacturer at varied settings and pressures.

DISCHARGE IN GALLONS PER MINUTE

NOZZLE PRESSURE IN POUNDS	NOZZLE DIAMETER IN INCHES					
	¾	1⅛	1¼	1½	1¾	2
40	106	237	292	422	575	752
50	118	265	326	472	648	841
70	140	313	386	558	761	994
100	168	374	461	667	909	1,189

This required fire flow in gallons per minute can be delivered through various sizes of hose, variously sized nozzle tips, and the use of varied pressures. Nozzle-tip sizes and nozzle pressures that can be handled readily and that at the same time can deliver a good effective stream have been calculated and field-tested. The standard flows from various sizes of nozzle tips using accepted nozzle pressures are shown in Table 11-1.

A "good" fire hydrant should be capable of delivering 250 gpm from each 2½-inch port. A "good" 1,000 gpm pumper in use will average 500 gpm fire flow. (Chapter 9 discusses this fire-fighting equipment in greater detail.)

By using the fire flow guidelines and hydraulic flow discharge tables, the number of estimated lines needed for most occupancies can be determined.

Example: A single-family residence of good construction in an area where exposures present no problem requires a fire flow potential of 500 gpm. This fire flow should be provided by a good hydrant flowing 250 gpm from each port. One pumping engine should be able to utilize this fire flow potential. A 1⅛-inch tip with 40- to 50-pound nozzle pressure will deliver 250 gpm into the fire building. The 500 gpm needed then will require two lines with 1⅛-inch tips from the one pumping engine.

Theory has provided another method of calculating the required flow for control and extinguishment of fire. A rate-of-flow formula has been derived, based on the amount of Btu that can be absorbed by water and the potential Btu that can be produced with one cubic foot of oxygen supporting combustion.

As previously discussed, we know that one pound of water going from 62°F to steam at 212°F will absorb 1,120 Btu (150 Btu plus latent heat of vaporization).

One gallon of water weighs 8.33 pounds. So, in one gallon of water, 1,120 Btu times 8.33, or 9,350 Btu per gallon, will be absorbed. Experiments have shown that one cubic foot of oxygen will support combustion yielding 535 Btu of heat; but air is only 21 percent oxygen. We also know that below 14 percent oxygen content, combustion will cease, so that in one cubic foot of air, there is only 7 percent oxygen available for combustion.

$$7\% \times 535 \text{ Btu} = 37 \text{ Btu per cubic foot}$$
$$200 \text{ cubic feet of air} \times 37 = 7,400 \text{ Btu}$$

This 7,400 Btu of heat should be absorbed by one gallon of water with a Btu-absorbing potential of 9,350 Btu. Therefore, if one gallon of water can absorb the heat potential of 200 cubic feet of area, the gpm flow for any area can be derived in the formula:

$$gpm = \frac{\text{cubic feet of area}}{200}$$

But evidence indicates this gpm flow should be applied in 30 seconds, so the formula must be amended to read:

$$gpm = \frac{\text{cubic feet of area}}{100}$$

Experience tells us that some of the required fire flow will fail to reach the seat of the fire and therefore will be wasted. To provide for this water loss and its effect on the fire, a multiplying safety factor is added. The gpm required by the formula is multiplied by 3 or 4, yielding a safety factor of three or four times the quantity needed.

[1]This formula was originally derived by F.W.Nelson and Keith Royer in studies made at Iowa State University.

Theoretically, then, the required fire flow can be calculated for any size of building. It is only necessary to determine the cubic-foot contents of the building and apply the formula. (The required gallon-per-minute flow then has to be allotted to lines that can surround the fire area and deliver the individual line flows simultaneously.)

A comparison may be made between the Insurance Service Office's guideline of 500 gpm for a single-family residence and the formula calculation. If a single-family residence is assumed to have between 1,200 and 1,600 square feet of area, then a factor of 3 and the formula would call for 290 to 390 gpm. A factor of 4 and the formula would require 385 to 510 gpm fire flow.

Ceiling height of the average residence is 8 feet
1,200 × 8 = 9,600 cubic feet ÷ 100 = 96 × 3 = 288
1,600 × 8 = 12,800 cubic feet ÷ 100 = 128 × 3 = 384

From a practical standpoint, the experienced fire officer can estimate the number of lines necessary for control from past experience. The theoretical and practical lines needed can be based on the standard of a 2½-inch line with a 1⅛-inch tip of 40- to 50-pound nozzle pressure delivering 250 gpm. The number of pumpers needed can be estimated. Hydrants in the immediate area of the fire building can be flow-tested to see whether the estimated fire flow is available.

BASIC ATTACK TEAM RESPONSIBILITIES

Strategical considerations of fire control and extinguishment have established guidelines to be followed in the placement of fire equipment at the fire scene. These basic attack positions are considered in the size-up and the ensuing tactical plan. In essence, the placement of equipment is a practical application of the theory of containment and confinement followed by extinguishment. The primary attack is normally from the "front" of the fire. It should be borne in mind, however, that this front may not be the front of the fire building, but instead may be the rear or the sides as well.

Engine Positions

The basic attack positions of the engine company (see Figure 11-1) are these:

1. A line to the inside of the building from the front of the fire

2. A line above the fire to cover upward exposure

3. A line to attack the fire from the rear

4. A line above the fire from the rear

5. A line at the fire side of the building—possibly into the adjoining building (point of vantage)

6. A line at the other side

7. Heavy streams where needed

8. In a sprinklered building, the placement of supplementary lines into the sprinkler siamese connection by the first or second responding engines — an imperative responsibility

Many fires may not require a two-front attack; an attack with one stream may be all that is necessary. In other cases, an inside attack may have to be ruled out immediately on arrival and a defensive line set up, with an attack on the fire being made by heavy stream appliances.

A precaution is in order. In using elevators to get to the fire floor, it is imperative that the elevator stop at the floor below, or several floors below, the fire floor. Many firefighters have lost their lives when trapped on the fire floor in an opened elevator that was made inoperative by smoke or fire conditions.

Good theory dictates that the initial fire attack should have two pumping engines in the basic attack team. One engine can then assume responsibility for the front of the fire and the crew of the second pumper can lead out necessary lines to the rear. The engine company that appears to have the best shot at the fire can then press the attack while the other company stands by with lines ready for any extension of the fire that may develop. The danger in a two-front attack is that each company may drive heat and smoke toward the other and thus retard the other's advance. When the fire is confined, the company that has the best chance of reaching the seat of the fire should press the attack.

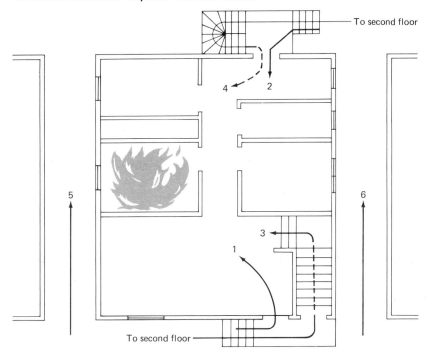

Figure 11-1 Basic engine positions to be covered in the fire attack. The front of the fire is the position covered first and, in application, may be a line to protect exposures. Other positions are covered in numerical sequence if the need exists.

There are certain other precautions that must be observed. In the desire to throw water on the fire itself, excessive lines are sometimes used. One line that is hitting fire can extinguish a large volume. Secondary lines would be better utilized in covering possible exposures, with firefighters checking those areas for possible communication, and opening the nozzle only if fire is discovered in the exposures. It is of little value to extinguish fire on the first floor only to see it break out in the attic space two floors above. On the other hand, it is pointless to cool down a possible exterior exposure with a line that could be better used on the seat of the fire. This line, serving a dual purpose, could protect the exposure and extinguish the fire.

Gaining control of the floor above the fire floor is one of the most difficult tasks for a fire department. Heat and smoke may have risen to fill this area, making it untenable. Control must be obtained by proper ventilation procedures and a line brought up to extinguish any fire found in the routine checking of all possible means of fire communication.

In general, a hand line of 1½ inches can be used in the interior fire attack. Lines that are 2½ inches or larger, either single or siamesed, will be used for hand lines or master-stream appliances in the outside attack. Again, caution must be observed. While it makes good sense to use the smaller line and less water on the interior fire, the larger line must be available for use when needed in the interior attack.

Individual department practices will usually determine the lead-out and evolution used, with the decision being based on the personnel and equipment available to the department.

Truck Duties

The ladder truck responding (and good practice states that a ladder truck should be in the basic attack team) will also be governed by basic attack positions and actions. They are:

1. Aerial to roof where needed
2. Laddering of building as needed
3. Forcible entry and ventilation at the front of the fire
4. Forcible entry and ventilation at the rear of the fire

These responsibilities will be carried out concurrently whenever possible. The engine attack team will be operating as a unified force. The ladder truck complement of men will by necessity be split up so that several phases of work can go on simultaneously. When a working fire requires the two engines to be operating their lines simultaneously, it can be readily seen that the hook-and-ladder company will be unable to aid both companies properly. If a salvage company or a squad company is included in the basic attack team, this unit can assume some of the truck work needed by the second engine company. If no further help is in the basic attack response, more help will have to be called if quick extinguishment is going to be obtained.

Ventilation

The saving of life of course is paramount. In most fires, there is no life hazard involved, so the primary task of the hook-and-ladder company becomes that of ventilation. Ventilation, in terms of the fire service, may be defined as the careful creation of vents in a burning structure so that noxious fumes and gases can be released to the atmosphere, to be replaced by fresh air. Ventilation will thus aid rescue operations, help in gaining fire control and extinguishment, and curtail smoke and fire damage.

The engine company is unable to press an interior attack where smoke and heat have reached temperatures above normal endurance. The importance of ventilation may be summarized thus:

1. It helps locate the fire. As smoke and heat are vented from the building, the fire can be more readily detected. The introduction of air to the building may supply oxygen to the fire seat, with the resulting flare-up indicating the location of the fire.

2. It aids in rescue work. The venting of the smoke, gases, and heat gives the trapped victim a new lease on life and the firefighter-rescuer a better chance to reach the victim.

3. It prevents explosion or back draft. Proper ventilation above the seat of the fire will release enough heat and smoke so that back-draft and explosion possibilities are virtually eliminated.

4. It eliminates the spread of fire by mushrooming smoke and buildup of heat.

5. It enables the firefighter more easily to advance to the seat of the fire.

6. It lessens the health hazard of smoke, heat, and gases.

7. It minimizes smoke, heat, and water damage.

Ventilation should be effected vertically over the fire wherever possible. This approach eliminates the possible back-draft explosion and allows the rising hot gases and smoke to escape from the fire building. This vertical ventilation can be accomplished by the opening of the roof or skylights and scuttleholes on the roof. Floor-to-floor ventilation may be accomplished by openings in the floor, stairways, or shafts.

The opening or breaking of windows is the fastest and easiest method of horizontal ventilation and ventilation that permits escape of heat and smoke from the fire floor. The opening of partitions and interior doors or breaching of exterior walls also permits horizontal ventilation of a particular fire floor.

There are inherent dangers in ventilation practices that are not carefully planned. The danger of back draft may occur where ventilation permits oxygen to reach the seat of the fire before some smoke and heat have been released. There is the danger of a free-burning condition when air supplies oxygen to the oxygen-starved fire. This hazard can be overcome when a charged line is standing by before ventilation proceeds.

Vent openings must be large enough to do the job. An 8- by 8-foot hole in a roof supplies twice the venting capabilities of two 4- by 4-foot openings.

All glass from broken windows must be removed, and curtains and shades as well. Where windows are opened from the inside, both the top and bottom sashes of double-hung windows should be opened. Casement and awning-type windows should be fully opened on adjoining or opposite walls, if possible.

Forcible entry is a possibility. Walls and ceilings must be opened up for possible fire travel. Much work is encompassed in the term "general truck duty." The truck man must be well drilled and disciplined so that, when necessary, he can operate as a unit of one in his myriad duties at a working fire. (See Figure 11-2.)

In some cases, the fire will have gained such headway that ventilation or the laddering of the building is not called for. In other instances, the fire may not be large enough to warrant extensive laddering or venting. The ladder company will then assist the engine company in the hose lead-out or perhaps will man a line or master-stream appliance.

COMMAND DUTIES

The basic attack team will be headed by a chief officer, responsible for the tactical solution of the fire problem. If the companies under his command are well trained in basic attack duties, his task will be an easy one.

Arriving on the fire scene as a unit, the members of the attack team will almost automatically commence the various tasks that

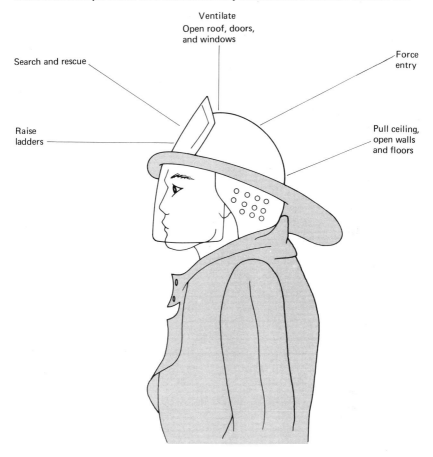

Figure 11-2 Duties of the ladder truck companies in the initial response. The many and varied duties of the truck personnel often demand that they work as individuals rather than as a unit.

Ventilate
Open roof, doors,
and windows

Search and rescue

Force
entry

Raise
ladders

Pull ceiling,
open walls
and floors

need doing in order of priority. This is as it should be. The chief officer cannot literally lead each firefighter by the hand to the task that needs to be done. All members of the well-trained company will know their role and position at the fire scene and will go to work. The chief officer can make a quick survey of the fire building and, through experience and his overall view of the fire scene, evaluate efforts and amend attack methods where necessary. Perhaps he orders the second line shut down so that the first line can advance. Perhaps he orders the opening of the roof halted as it apparently is not necessary now. In some cases, he may have to order more equipment to the scene.

Having observed the extent of fire, he can now intelligently direct the extra companies to the most advantageous position. His own position is at the front of the fire so that units can report to him, but the position must be a flexible one. While his aide is manning the command center, he will tour the fire building and surroundings to see what progress is being made or what needs doing. It is not his responsibility to direct the engine company working on the seat of the fire, but on occasion he will be in the thick of the battle with the company to see what progress is being made, or possibly to change tactics because of his visual observation of the overall fire scene. He cannot adapt his plan to current conditions unless he observes the work in progress and the results of the efforts being expended.

Where more help is called for and other chief officers respond, a command will be established at the rear and at the sides of the fire building so that a concentrated view and command, under one commanding chief, can be obtained on all sides of the fire. (See Figure 11-3.)

The chief officer at the fire scene has an additional responsibility when another chief officer or fire alarm center is a part of the municipality's fire protection system. That responsibility consists of informing the fire alarm center of the nature of the fire and its possible extent. This report will alert the fire alarm center of possible needs at the fire or of equipment available for a second fire. The following information should be relayed to the fire alarm center as quickly as possible after arrival at the fire.

1. Size of fire building and construction

2. Occupancy

3. Extent of fire — area involved

4. Life hazard involved

5. Exposure potential, both exterior and interior

6. Extra equipment needed, either fire department or other

A progress report should be given from time to time as the attack on the fire continues, stating conditions of the fire and future possibilities.

Figure 11-3 Command positions at the large structural fire. Proper fire ground coverage requires that someone be delegated with authority for instant command decisions on all sides of the fire under the leadership and guidance of an overall fire ground commander. When the fire building can be entered, the need for unit area commanders is less. Position A (at the front of the fire) is the command center of the fire ground commander.

If there is any doubt as to the confinement and extinguishment of the fire, the command officer should call for more help rather than wait until the moment of need arises before asking for aid.

It will also be the chief's responsibility to get companies back into service when no longer needed so that the company's own response area can be again protected. If advisable, he can keep these companies on the fire scene doing salvage work or aiding other companies, but in service for another alarm.

In summation, fire tactics is putting the experience of other fires, the training, and the knowledge of drill evolutions into a practical plan of action warranted by the fire as seen on arrival at the fire scene. Each company, each person in that company, must know his role in the overall tactical plan and be ready to carry it out. The going may be rough for a short period of time, but an esprit de corps, the satisfaction and pride in the knowledge that a fire was well stopped, inspires the teamwork that in truth culminates in the accomplishment of a fire well stopped.

Small departments must depend on volunteers, the recall of paid men, and mutual aid pacts for extra manpower and equipment. Yet there will always be times, no matter how large the department, when not enough men and equipment are on the fire scene to efficiently contain and extinguish the fire.

The fire officer must learn to make do with what he has. If he cannot press an attack, he will confine the fire to an area he can control. Total extinguishment will take longer, but he will have stopped the fire.

SUMMARY

The fire service, of necessity, must be ready to initiate a plan of action immediately upon arrival at the fire ground. The strategy planned to combat fire will include the acquisition of necessary equipment and the formulation of drill evolutions using this equipment in tactical maneuvers. At the fire scene, the proper equipment and evolution for the particular incident will be used, changes being made as necessary.

Theory, experimentation, and practical experience at the fire ground offer guidelines to strategical considerations of the number and placement of lines.

Basic positions must be covered to confine and control the fire.

1. A line to the inside from the front of the fire
2. A line above the fire from the front (vertical exposure)
3. A line to attack the fire from the rear
4. A line above the fire at the rear
5. Lines at the side exposures of the fire
6. Master streams where needed

In essence, this is an attack from front and rear — over and under the fire; specific fire conditions will dictate the order of placement.

The responding ladder truck will also have basic duties and positions to be covered.

1. Aerial to the roof if needed

2. Laddering of the building where needed

3. Forcible entry and ventilation of the front — interior and exterior

4. Forcible entry and ventilation of the rear — interior and exterior

The command officer will maintain a command center at the front of the fire but will also tour the fire building to evaluate and amend attack procedures in use.

If the size of the fire warrants, there will be subcommanders at other positions of the fire.

DISCUSSION TOPICS

1. Why is time so important to the firefighter?

2. Estimate the theoretical required fire flow for various sizes of buildings, using the formula

$$\text{gpm} = \frac{\text{cubic feet of area} \times 3}{100}$$

3. What basic positions must be covered by the engine company to cover a fire properly?

4. What are the principal truck duties?

5. Why is ventilation so important?

6. Discuss command responsibilities at the fire scene.

RESEARCH PROJECTS

1. Select several residential building locations in the local area. Plot these buildings on a sheet of paper. Using the Insurance Service Office grading schedule of fire flow recommendations, sketch in the number and size of lines and their location for coverage of the buildings, and spot hydrants to be used in estimated fire flow.

2. Using the same buildings, estimate their cubic contents. Using the formula

$$\text{gpm} = \frac{\text{cubic feet of area}}{100}$$

calçulate the ideal amount of water flow needed. Multiply by a factor of 3, and also 4, to determine more realistic water-flow requirements. Compare with ISO requirements for the same buildings.

3. Using the same buildings, mark in the required laddering of buildings.

4. Select one mercantile or industrial building. Assuming the building is well involved in fire, dete·.nine the number, size, and location of lines estimated to be necessary for confinement and extinguishment. Do the same for ladders. Determine command and subcommand positions.

FURTHER READINGS

The publications listed in Chapter 10 are equally valuable to Chapter 11.

The complex of industry found in our society offers many problems to today's firefighter that the firefighter of the past did not have to face. The increasingly great use of chemicals in industry presents a sophisticated challenge. Many newly developed products are without proper safeguards; no one seems to have a real understanding of the reaction that may take place under fire conditions. At best, the firefighter can come to know the characteristics of only those most commonly found.

For the others, he depends somewhat upon the information that can be given him by plant personnel, engineers, chemists, or maintenance men when the need arises. This information often leaves much to be desired.

There are some ready reference sources for referral in those emergencies where no quick or easy solution is apparent. Guidelines for hazardous materials and the characteristics of many individual or trade name chemicals, liquids, gases, and volatile solids have been published by many sources. A partial list follows:

NFPA Handbook No. 325M, *Fire Hazard Properties of Flammable Liquids, Gases, Volatile Solids*

NFPA Handbook No. 49, *Hazardous Chemical Data*

Manufacturing Chemists' Association, *MCA Chemical Card Manual*

American Insurance Association, Technical Survey No. 3, *Hazard Survey of the Chemical and Allied Industries*

NFPA Handbook No. 704M, *Identification of the Fire Hazards of Materials*

The fire service must have access to these or other reference

CHEMICAL HAZARDS

sources that give some information on the characteristics of the less common chemicals encountered in some incidents.[1]

The hazard need not be confined to a plant process. Many of the products that possess dangerous characteristics are regular articles of commerce. Transportation incidents involving fire or explosion are described almost daily in the country's newspapers. The incident may occur in the large metropolitan area or in a small hamlet unprepared for the magnitude of the catastrophe.

DEPARTMENT OF TRANSPORTATION

To reduce personal injury, loss of life, and damage to property, Congress has directed the Department of Transportation (DOT) to formulate regulations governing the transportation of certain hazardous materials. Eight different hazard classifications have been established. They are:

1. Explosives. Any chemical compound whose purpose is to explode
a. Explosive of maximum hazard (TNT, etc.)
b. Materials capable of rapid combustion rather than detonation
c. Manufactured articles with a class A or B component in a restricted amount (flares, etc.)

2. Poisonous articles
a. Substances that are extremely dangerous even in small amounts
b. Materials other than those in (a) and (c) that are hazardous to health through toxicity via breathing or skin contact
c. Tear gas or irritants

3. Flammable liquids

4. Flammable solids, other than explosives, that may become fire hazards by friction, moisture, heat, or spontaneous ignition

5. Oxidizing materials that contribute to combustion (chlorates, nitrates, permanganate peroxide, etc.)

6. Corrosive liquids that may cause skin damage or cause or contribute to fire in other materials

[1]The above manuals or similar reference sources should be in the private library of the fire service officer.

7. Compressed gases

8. Radioactive materials

For substances in transit, these hazard classifications are identified by various placards which indicate only the broad classification of the commodity being carried. There is a definite need for a classification system that will not only place the commodity within its proper hazard class but will also give some indication of its particular hazard potential and methods and means of combating the fire or hazards present in the emergency.

To that end, the Division of Hazardous Materials in the Department of Transportation is considering a new marking system, a two-digit coding known as the HI System. The placard card or label will refer, through the two-number system, to specific information on the material in a booklet to be published by the Transportation Department.

It was some sixty years ago that the railroads and the Bureau of Explosives requested Congress to enact laws empowering the Interstate Commerce Commission to regulate the transportation of hazardous materials. Several years ago, this power to regulate was transferred from the ICC to the Department of Transportation. The DOT regulations apply to any type of common carrier as well as to railroads.

There appear to be ample regulations at the present time, but accidents do occur. Collisions and derailments are known to have happened. When fire is present, it is usually the result of the accident itself. Most often, the fire involves tank vehicles carrying flammable liquids.

Quite frequently the exact chemical and physical properties of the shipment are known only to the producer. There should be a better flow of information between the producer, the carrier, and the fire department.

Too often, the responding fire department can only "protect the exposures" in those cases where explosion and fire have occurred. In some cases it can establish a zone of peril because of the possibility of fire, explosion, or leakage of hazardous gases. It can warn persons in the imperiled area and see that they are safely removed. Some rescue work may be necessary. Many times, because of the extent and nature of the incident, nothing will really be accomplished unless a "disaster plan," encompassing perhaps several adjoining towns and necessary adjuncts such as hospitals and other services, has been established and rehearsed.

Efforts may be concentrated on the removal of explosives and hazardous materials with known characteristics. Those with unknown qualities may have to await disposal until a plant specialist who knows the product can assume control or advise on the operation. Also, a derailment always brings the possibility of the inadvertent mixture of the contents of several cars and an unknown or unexpected reaction.

CLASSIFICATION OF HAZARDOUS MATERIALS — NFPA Handbook No. 704M

A system that has been established for the fast identification of the hazards of various materials is detailed in NFPA Handbook No. 704M. The International Association of Fire Chiefs recommended that this system be adopted by the Department of Transportation for the marking of hazardous materials in transit. Not only does this system indicate the fire hazard, but also quickly catalogs the health and reactivity of the material. This information is given on a diamond-shaped placard that is divided into four smaller diamonds. Each small diamond is color-coded to identify the specific hazard cataloged.

Blue (left) diamond signifies the health hazard.

Red (upper) diamond signifies the fire or flammability hazard.

Yellow (right) diamond signifies the reactivity or instability characteristics of the hazard.

White (lower) diamond gives special information on the particular material. In most cases, this diamond will be blank. The usual indications here are, for example, that water should be used only with caution, or that the material is radioactive. The specific hazard potential is assigned a numeral of 0, 1, 2, 3, or 4, denoting the degree of hazard. The numeral 0 shows hazard potential or very little. The numeral 4 marks a very dangerous hazard.

Figure 12-1 gives more specific information on the degree of danger.

The plant that employs this marking system in areas where hazardous chemicals, gases, or materials are present offers a means of quick comprehension of the particular hazard. The diamond placard and its numerals must be large enough so that

Figure 12-1   Chart used in the classification of the hazards of materials, as recommended in NFPA Handbook No. 704M.

IDENTIFICATION OF THE FIRE HAZARDS OF MATERIALS

DEGREE OF HAZARD

Identification of Health Hazard Color Code: BLUE	Identification of Flammability Color Code: RED	Identification of Reactivity (Stability) Color Code: YELLOW
Type of Possible Injury	Susceptibility to Release of Energy	Susceptibility of Materials to Burning
Signal	Signal	Signal
4 — Materials which, on very short exposure, could cause death or major residual injury even though prompt medical treatment were given.	4 — Materials which will rapidly or completely vaporize at atmospheric pressure and normal ambient temperature, or which are readily dispersed in air and which will burn readily.	4 — Materials which are readily capable of detonation or of explosive decomposition or reaction at normal temperatures and pressures.
3 — Materials which, on short exposure, could cause serious temporary or residual injury even though prompt medical treatment were given.	3 — Liquids and solids that can be ignited under almost all ambient temperature conditions.	3 — Materials which are capable of detonation but require a strong initiating source, and which must be heated under confinement before initiation.
2 — Materials which, on intense or continued exposure, could cause temporary incapacitation or possible residual injury unless prompt medical treatment is given.	2 — Materials that must be moderately heated or exposed to relatively high ambient temperatures before ignition can occur.	2 — Materials which readily undergo violent chemical change.
1 — Materials which, on exposure, would cause irritation but only minor residual injury even if no treatment is given.	1 — Materials that must be preheated before ignition can occur.	1 — Materials which are normally stable, but which can become unstable in combination with other common materials or at elevated temperatures and pressures.
0 — Materials which, on exposure under fire conditions, would offer no hazard beyond that of ordinary combustible material.	0 — Materials that will not burn.	0 — Materials which are normally stable.

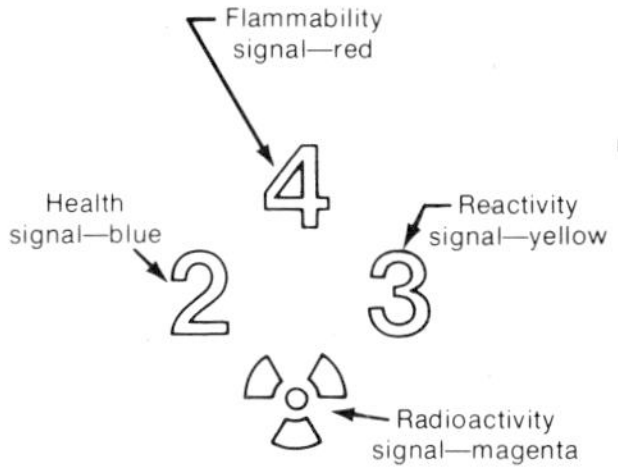

Flammability signal—red

Health signal—blue

Reactivity signal—yellow

Radioactivity signal—magenta

For Use Where White Background Is Not Necessary

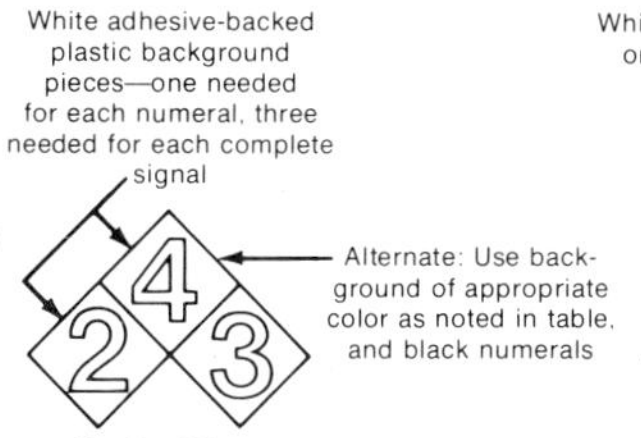

White adhesive-backed plastic background pieces—one needed for each numeral, three needed for each complete signal

Alternate: Use background of appropriate color as noted in table, and black numerals

For Use Where Background Is Used with Numerals Made from Adhesive-Backed Plastic

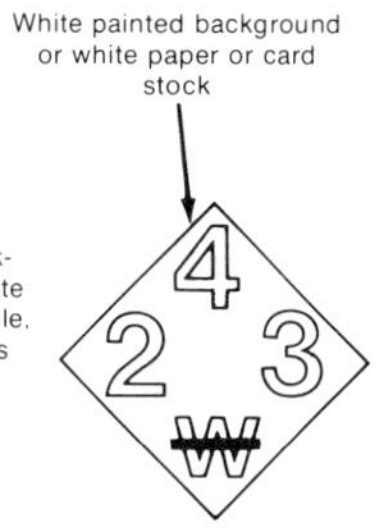

White painted background or white paper or card stock

For Use Where White Background Is Used with Painted Numerals or for Use When Signal Is in the Form of Sign or Placard

Distance at Which Signals Must Be Legible (ft)	Size of Signals Required (in.)
50	1
75	2
100	3
200	4
300	6

Courtesy of the National Fire Protection Association.

they can be seen and read at a distance. If no further information is known or obtainable on the particular hazard, the firefighter can proceed with reasonable assurance and observe necessary precautions indicated by the numeral classification. (Figure 12-2 shows one example.)

This classification system is used in NFPA Handbook No. 49, *Hazardous Chemicals Data,* and NFPA Handbook No. 325M, *Fire Hazard Properties of Flammable Liquids, Gases, and Volatile Solids,* along with more specific information on the particular chemical, liquid, gas, or solid.

PROPERTIES OF FLAMMABLE LIQUIDS AND GASES

The firefighter is frequently exposed to the fire and explosive hazard of flammable liquids and gases. To understand the hazard of a certain liquid or solid, he must know the various properties that the hazard possesses.

Figure 12-2   A model installation employing the hazard classification recommended in NFPA Handbook No. 704M. Even without specific knowledge of the properties of the classified substance, the firefighter is immediately aware of the potential hazard he is facing.

Flashpoint

A liquid will not burn. It is the vapors that are given off when the liquid is heated that burn. When a liquid is being heated its flashpoint is reached at the temperature at which enough vapor is given off to form an ignitable mixture with the oxygen present. The vapors may not, however, continue to burn unless the temperature rises several more degrees. At this point, the fire point is reached and sufficient vapors will be released to maintain continuous combustion.

Ignition Temperature

A substance, that is, any liquid, gas, or solid, can be ignited if a source of ignition is applied that is hot enough and applied long enough. Up to a point in temperature peculiar to the particular substance, the substance will stop burning if the source of ignition is removed. When that particular temperature is reached, the substance will continue to burn of itself and the fire will be self-sustaining. This is the ignition temperature of that substance.

Flammable Explosive Limits

In addition to the need for the flashpoint temperature of flammable gases to be reached before ignition occurs, there must be a minimum percentage of the vapors in the air or oxygen for propagation of flame to occur. Below the lowest percentage, the mixture is too "lean" to burn. Above a saturation percentage of vapor to air, propagation of flame will also not take place — the mixture is too "rich" to burn. We are familiar with both terms through the mechanics of the gasoline engine. These two percentage figures are the upper and lower flammable explosive limits.

Flammable Explosive Range

The spread between the upper and lower limits of the propagation of flame of a particular substance is known as the flammable explosive range.

Specific Gravity

The specific gravity of a particular substance is the ratio of the weight of the substance to the weight of another given substance, both having the same volume. Specific gravity, as the term is most commonly used, is the ratio of the particular substance in weight to the weight of an equal volume of water. The importance of this fact to the firefighter can be readily seen. A substance with a specific gravity greater than 1, or heavier than water, will sink in the water, and if on fire, will be extinguished. A substance with a specific gravity less than 1, or lighter than water, will float on the water, and if on fire, will continue to burn and perhaps be spread out over a greater area.

Vapor Density

Vapor density is similar to specific gravity but is a characteristic of vapors or gases. It is the ratio of the weight of a volume of gas to the weight of an equal volume of air. Air is taken as 1. A gas less than 1 is therefore lighter than air and will rise. A gas over 1 is heavier than air and will settle to the ground in low pockets, presenting a severe potential problem to the firefighter. Gases heavier than air have been known to travel as far as 300 feet along the ground to a source of ignition.

Boiling Point

The boiling point of a liquid is the temperature of the liquid at sea level when its vapor pressure is equal to the standard sea-level atmospheric pressure of 14.7 pounds per square inch; it is given in degrees Fahrenheit. This boiling-point temperature indicates the volatility and consequent hazard of the liquid; the lower the boiling point the greater the volatility.

Viscosity

The viscosity of a liquid is a measurement of its ability to flow freely. The difference in the rate of flow of two liquids is due to the differing internal fluid resistance of the two liquids. A heavier, thicker liquid will be more viscous and will flow more slowly. A freely flowing liquid, such as gasoline, can spread fire over a wide area. A heavy oil will spread slowly and fire can be confined to a smaller area.

Water Solubility (Miscibility)

This quality indicates the characteristics of a liquid to mix with water. Liquids that are miscible will have their flashpoint and fire point temperatures raised as water is added to the solution.

FIRE SUPPRESSION AND FLAMMABLE LIQUIDS AND GASES

The properties of liquids and gases, just described, can be used in determining the type of extinguishing agent to be used.

Flammable liquid and gas fires are classified as class B fires. Foam, dry chemicals, carbon dioxide, and the vaporizing liquids, such as the halogens, appear to be the best extinguishing mediums, perhaps in a one-two punch of dry chemicals followed by foam.

Water spray can be used with a great deal of success in liquids having a flashpoint over 100 °F. In the liquids whose flashpoints are above 212 °F, a great deal of frothing may occur when water fog or spray is used.

A sprinkler system, nozzle spray, or solid stream is still the best tool of the firefighter on storage tanks or drums containing flammable liquids or gases. Recall the expansion of liquids and gases when subjected to heat. This expansion places a greater pressure on the tank — perhaps sufficient pressure to rupture the container, permitting the volatile liquid or gas to explode and burn. Where tanks or drums of flammable liquids and gases are within a fire area and subject to temperature rise, the firefighter's primary responsibility is to cover the tanks with a water stream to prevent the buildup of heat and the resulting increase in pressure.

Other cases of flammable liquid or gas fires are often caused by open valves or broken feed lines. No extinguishing agent that cools or smothers is necessary when the firefighter can approach the tank, perhaps under a water curtain, and shut off the valve feeding the ruptured line or open pipe, thus causing a separation of fuel from the ignition source.

Where tanks are punctured or shutoff valves destroyed and a flammable gas is burning, the firefighter should not extinguish the flame unless a positive means of stopping the gas supply is available. If he does extinguish the fire and the gas continues to escape, he is subjecting the area to an explosion when the gas concentration builds up and a source of ignition is present.

NFPA Handbook No. 325M, *Fire Hazard Properties of Flammable Liquids, Gases, and Volatile Solids* (1969 edition), lists more than

Table 12-1 Some Flammable Substances and Their Fire Hazard Properties

		PROPERTIES				DEGREE OF HAZARD		
SUBSTANCE	FLASHPOINT	EXPLOSIVE RANGE	IGNITION TEMPERATURE °F	SPECIFIC GRAVITY	VAPOR DENSITY	HEALTH	FIRE	STABILITY
Chlorine	Does not burn but supports combustion			1.50	2.50	3	0	1
Acetylene	Gas	2.5–82%	581°	—	.90	1	4	3
Ammonia	Gas	16–25%	1204°	—	.58	3	1	0
Carbon bisulfide	−22°F	1.3–44%	212°	1.30	2.64	2	3	0
Hydrogen	Gas	4.0–75%	1085°	—	0.07	0	4	0
Propane	Gas	2.2–9.5%	842°	—	1.56	1	4	0
Butane	Gas	1.8–8.4%	761°	.58	2.01	1	4	0
Methane	Gas	5.0–15%	1004°	—	1.34	1	4	0
Gasoline	−45°F	1.4–7.6%	500°–824°	.75	3.0–4.0	1	3	0
Ether	−49°F	1.9–36%	356°	.71	2.56	2	4	0

1,300 substances, too many for the firefighter to learn. However, there are several that the firefighter is most apt to come in contact with and that exemplify the various characteristics and their potential. They are listed in Table 12-1.

Chlorine

Chlorine is a noncombustible gas, but it does aid combustion of many other combustibles, as does air. It can form explosive mixtures with the vapors of certain flammable gases and liquids. A very poisonous gas, its greatest hazard is its toxicity. Therefore, its health hazard classification is 3.

Its specific gravity and vapor density intensify the hazard in this gas. In a liquid state, it can cause serious burns. As a gas, it will tend to settle in low areas and remain for a long time as the diffusion rate is very slow. These hazards create the need for evacuation of the area involved in a chlorine leak and the need for full protective clothing and masks. Chlorine leaks in storage tanks are caused by a corrosive action that is only increased by the application of water. Leaks must be plugged by special emergency kits usually on the premises in the correct size for the type of tank in use. Water can be used, however, on tanks that are exposed to heat from a fire and therefore in danger of gas expansion.

Acetylene

Acetylene is a commercial gas generated by the action of water on calcium carbide. It is a colorless, flammable gas with a wide explosive range—2.5 percent to 81 percent. Ignition temperature is below that of most combustibles. Its vapor density of .90 shows that it is lighter than air and will rise and be dissipated. Thus it does not present the hazards that the heavier-than-air gases do.

If an acetylene tank is burning through an open valve, the firefighter will attempt to shut the valve off. It the tank is ruptured or cannot be shut off, it is usually better to let the gas burn out. Water can be used to cool the tank to prevent gas expansion. The degree of fire hazard, 4, indicates the great fire potential.

Ammonia

The degree-of-hazard classification indicates that ammonia offers a serious health hazard, rated 3, to the firefighter and requires full protective clothing. The fire hazard is only rated 1; flammable limits range from 16 to 25 percent, but a rather high ignition temperature, 1204°F, is required.

Ammonia has an extremely pungent odor and is extremely irritating to the eyes, mucous membranes, and skin. The victim is forced to flee the area where possible. Ammonia is lighter than air, rises, and the gas tends to be absorbed by a water spray. Plenty of water should be used on the ammonia gas but not on the ammonia in a liquid state.

The boiling point of liquid ammonia is 628°F, which will freeze skin tissue. Ammonia subjected to moisture has a caustic action on the skin and also reacts with copper, zinc, and many alloys.

Carbon Bisulfide

Carbon bisulfide is a liquid that carries a life hazard rating of 2, being toxic by skin contact or ingestion. Other characteristics present more severe hazards. The fire classification is 3. The substance has a flashpoint of 22°F, a wide explosive range of 1.3 to 44 percent, and the low ignition temperature of 212°F. This ignition temperature can be reached through physical contact with a lighted bulb or heated surface. The vapor is heavier than air and settles low, possibly traveling a distance to the source of low ignition temperature. Dry chemicals and carbon dioxide are recommended extinguishing agents, with water as the coolant for tanks or containers.

These five substances are examples of liquids and gases that give some indication of their hazard by their characteristics. They also present health or reactivity hazards as shown by their degree-of-hazard classification.

The following liquids and gases present in the main a fire hazard only—a fire hazard as indicated by the characteristics.

Propane and Butane

These gases are two of the liquefied petroleum gases. They are placed in tanks under pressure until they liquefy. Upon being released from the container, the liquefied petroleum gas returns to the gaseous state. Being heavier than air, they tend to settle in low pockets, dissipate slowly, and present a serious explosion hazard. They are the bottled gas used where natural gas for cooking is not available, or they may be mixed with the natural gas as a supplement. Propane and butane also are commercially used as a fuel source similar to gasoline or diesel fuel in an automotive engine.

Protective measures once again are separation of the fuel from the fire by shutting off the source of supply whenever possible. Burning gas should be allowed to burn until it can be shut off. Tanks subject to heat should be cooled. The firefighter will use water spray on escaping gas so that the air-gas lower explosive limit will be more difficult to reach. He will shut off ignition sources in the path of gas travel and be watchful of low pocket areas.

Gasoline

Gasoline is one of the most common hazards that the firefighter faces. The large amount consumed by the automobile brings this hazard home to everyone, but has also resulted in many safety features being introduced by the gasoline industry. Automatic shutoffs on fuel pipes and tanks, grounding devices, and other safeguards impart a degree of safety, but the inherent hazard and human error or mechanical failure are always present.

The hazard is indicated by the low flashpoint—45°F—indicating that flammable vapors are for practical purposes always being given off. Gasoline has a fairly narrow explosive range but a low ignition temperature of 500° to 850°F, depending on the grade. Again, it is a gas whose vapors are heavier than air—much heavier—and they settle in low pockets. The liquid, being lighter than water, will float on the water, making water almost useless as an extinguishing agent, although in some cases a water spray may effect smothering extinguishment. Water, of course, is most effective in tank cooling where tanks are subject to undue heat. The most effective extinguishment is by dry chemical, foam, or carbon dioxide.

Ether (Ethyl Ether)

Ether is a gas more hazardous than gasoline. It has a lower flashpoint (49°F), a wider explosive range (1.9 to 36 percent), and a lower ignition temperature (356°F). Its vapor density is also heavier than air, giving it the potential of traveling to a source of ignition. Ether does carry a health hazard of 2 because of its anesthetic ability to induce unconsciousness. Self-contained breathing equipment is thus needed in the gaseous area.

Dry chemical, foam, or carbon dioxide is used as the extinguisher, with water used to cool the fire-exposed container.

Unknown Substances

These examples of more common hazards and their characteristics should become familiar to the firefighter. The characteristics of the unknown liquid or gas will be more readily understood through an understanding of the ratings and the interrelationship of characteristics of the more common hazards.

An understanding of these exemplary substances also leads to a method of classifying other materials. If a liquid X and its characteristics are not readily known but can be said to be similar to ether, the individual who is familiar with ether can readily understand the potential hazard of liquid X. This is in essence the Underwriters' Laboratories classification for grading relative flammability. Liquids and gases are placed in a classification system that identifies the liquid or gas with a more common liquid and its characteristics. The system is shown in Table 12-2.

Liquids can also be categorized in a class rating that is based on the flashpoint of the liquid, as shown in Table 12-3. With this basic understanding of the fire potential of liquids, gases, and volatile solids, firefighters can readily understand the information offered in the references available to them when they need information on a particular substance.

Hazardous Materials in Industry

The fire potential of an industry or an industrial process using the chemicals that are classed as hazardous will of course depend on the type of operation, process, or chemicals in use. There is usually the greater possibility of an explosion, or of a much more rapid release of energy, than in the usual class A fire. The causes of the hazards and the steps to be taken to guard against the dangers in chemicals or processes that have been around for some time are usually known to industry. And industry does not want the fire any more than the firefighter who must face the risk. In most cases, the fire is caused by carelessness, poor maintenance, lack of supervision, or sloppy housekeeping.

The manager of a well-planned industrial plant or manufacturing process will have considered the occupancy hazard in relation to

Table 12-2 Relative Flammability, as Given in the Underwriters' Laboratories Classification

By this system the firefighter can relate the potential hazard of a relatively unknown substance with the hazard of a familiar substance.

CLASS	FLAMMABILITY
Ether	100
Gasoline	90 - 100
Alcohol (ethyl)	60 - 70
Kerosene	30 - 40
Paraffin oil	10 - 20

Source: Courtesy of Underwriters' Laboratories.

the type of construction. A reliable water supply and good outside fire protection are mandatory. Areas will have been divided for horizontal and vertical protection, and automatic sprinklers and fire detection systems will have been installed—perhaps some form of automatic fire protection employing CO_2, dry chemical, or foam will protect the chemical or the process.

The hazardous material must be segregated from other processes. In particular, incompatible chemicals must be kept apart.

The use of water on molten metals or molten minerals must be avoided if a violent explosion is to be prevented; many of the alkali metals, sodium, lithium, and potassium, are particularly reactive with water. The hydrogen and oxygen elements of water may undergo chemical reaction with the molten metal, leading to an explosion. Released oxygen may contribute to increased combustion. A rapid formation of steam may in itself be of explosive force.

The potential danger is always present in explosives, poisons, flammable liquids, flammable solids, oxidizing agents, corrosive liquids, compressed gases, and radioactive materials. A reexamina-

Table 12-3 Flashpoint Classification of Flammable Liquids

CLASS	FLASHPOINT*	BOILING POINT
1	Below 100°F	
1A	Below 73°F	Below 100°F
1B	Below 73°F	100°F or above
1C	73°F to 100°F	
2	100°F to 140°F	

*Liquids with a flashpoint above 140°F are classed as combustible liquids.

tion of the hazard classification makes this all too clear. The hazard need not be that of fire only, but also may present dangers to health and safety.

The problems that may be encountered only emphasize the need for company preplanning inspections to familiarize the firefighter with the type of occupancy and specific process. The need is apparent also for a ready reference source, such as NFPA Handbook No. 49 or NFPA Handbook No. 325M, to be at hand. A system of hazard identification, such as that described in NFPA Handbook No. 704M, is also a must.

RADIOACTIVE MATERIALS

Radioactive materials are now being used in the medical, research, and industrial fields. The fire hazard of the radioactive materials is the same as if the material were not radioactive, the particular hazard being in the alpha, beta, and gamma rays being emitted. The rays cannot be visibly seen, nor do they give off an odor; therefore, the victim can be in a dangerous concentration without realizing it.

The dose rate of emitted rays is measured in roentgens per hour. The term roentgen measures radiation in the air. The following table gives the effects on the human body of various total exposures to radiation in roentgen levels of emission.

ROENTGEN DOSAGE	EFFECTS
25	No detectable change
50	Temporary changes in blood
100	Radiation sickness, nausea, and vomiting
200	Kills 25 percent of persons exposed
400	Kills 50 percent of persons exposed
600	Kills all persons exposed

These figures are based on total body exposure for a short period (24 hours or less).

Radiation hazards can attack the body externally, through the skin, or internally, through ingestion. It is therefore imperative that the firefighter exposed to radiation hazards wear full protective clothing and use self-contained breathing equipment.

Because the dosage is measured in roentgens *per hour,* it is vital that each firefighter spend a minimum amount of time in the radiation area; the stronger the radiation dose, the less the time that can be spent.

Of greater importance is the distance to the radiation source. The dose rate falls off inversely as the square of the distance from the radiation source. For example, a 1,000-roentgen dose at 1 foot of distance will be only 111 roentgens (r) at 3 feet. Mathematically, 3 times 3 equals 9; one-ninth of 1,000 equals 111. At 5 feet, the dosage rate will be 40 roentgens. (Or, 5 times 5 equals 25; one twenty-fifths of 1,000 is 40.)

Another protection against the radiation hazard is by shielding the radiation source with a heavy, dense material such as lead. This is not feasible in firefighting, but it is the method of protection in storage and shipment. The radioactive material is usually shipped or stored in a lead case of necessary thickness to shield the amount of radiation being emitted.

A cleanup area must be maintained near the scene so that emission particles are not carried to other portions of the building. Approximately 75 percent of the emission particles that are on tools, equipment, and clothing can be flushed off with a water hose at the scene. The greatest danger is from the radioactive materials ingested into the body. They are not easily excreted and tend to travel to radiosensitive body organs.

SUMMARY

The storage of chemicals, new processes, and new materials offers many potential problems to the firefighter. Information is often lacking or unobtainable on particular hazards when the emergency arises.

The firefighter must become familiar with the characteristics of the more common chemical hazards. For others, he must have current guidelines available on trade-name flammable liquids, gases, and volatile solids.

Booklets have been published by the National Fire Protection Association, American Insurance Association, and specific product associations or companies listing the fire hazard properties of many substances.

These hazards may also be encountered in transportation incidents involving fire or wreck. It is the responsibility of the U.S.

Department of Transportation to formulate regulations governing the transportation of hazardous materials.

A need for a method of fast identification of the potential hazard both in transit and on-site location is apparent. The National Fire Protection Association has devised a means of ready identification of the potential hazards of a substance by placing it in one of three categories: (1) health; (2) flammability; and (3) reactivity.

In essence, the substance is assigned a numeral classification in each category—0 to 4; 0 represents no hazard or little hazard, whereas 4 represents a very severe hazard.

Flammable liquids, gases, and volatile solids all possess certain properties. An understanding of these properties leads to an understanding of the hazard of the particular liquid, gas, or volatile solid. These properties are:

1. Flashpoint and firepoint

2. Ignition temperature

3. Flammable explosive limits and range

4. Specific gravity

5. Vapor density

6. Boiling point

7. Water solubility

Among the flammable liquids and gases that may be frequently encountered are chlorine, acetylene, ammonia, carbon bisulfide, hydrogen, propane, butane, methane, gasoline, and ether. The properties of these liquids should become familiar to the firefighter.

Radioactive materials are particularly hazardous not only because of the lethal effects of the emission but also because the victim is unaware that he or she is in a dangerous concentration. Radiation can affect the victim externally through the skin or internally through ingestion of radioactive particles. Time (minimum exposure to the hazard), distance (increased distance from the hazard), and shielding (protection by thick concrete walls or lead shielding) are the best protections against excessive dosage of radioactive emissions.

DISCUSSION TOPICS

1. Discuss the relative merits of the Transportation Department's HI classification system and the classification system in NFPA's Handbook No. 704M.

2. Compare flashpoint and ignition temperature.

3. What is meant by flammable explosive limits?

4. Why is the specific gravity of a substance important to the firefighter? Why is vapor density also important?

5. What are the characteristics of carbon bisulfide that make it particularly hazardous?

RESEARCH PROJECTS

1. Survey the local area for industries using flammable liquids, gases, and volatile solids. List these flammable substances.

2. Using NFPA Handbook No. 49, *Hazardous Chemicals Data,* or NFPA Handbook No. 325M, *Properties of Flammable Liquids, Gases, and Volatile Solids,* list the properties of the substances found in project 1.

3. Assume it is your responsibility as plant safety engineer to placard the hazards of ten substances in use in the plant. Using NFPA Handbook No. 704M as a guideline, classify the hazard of ten chemicals assumed used in the plant.

4. Research American Insurance Association bulletins for accounts of incidents involving transportation accidents.

5. Check recent and current newspapers and magazines for stories of fires or explosions involving chemicals on site locations or in transit. List these stories and give a brief account of each event.

FURTHER READINGS

Manufacturing Chemists' Association, *MCA Chem Card Manual,* Washington.

National Board of Fire Underwriters, NBFU Research Report No. 1, *Fire Hazards of the Plastic Manufacturing and Fabricating Industries,* New York.

American Insurance Association, Technical Survey No. 3, *Hazard Survey of the Chemical and Allied Industries,* New York, 1968.

National Fire Protection Association, Boston.

Pamphlet Handbook No. 30, *Flammable and Combustible Liquids Code.*

Pamphlet Handbook No. 704M, *Recommended System for the Identification of the Fire Hazards of Materials.*

Pamphlet Handbook No. 49, *Hazardous Chemicals Data.*

Pamphlet No. 491M, *Manual of Hazardous Chemical Reactions.*

Pamphlet No. 325M, *Fire Hazard Properties of Flammable Liquids, Gases, Volatile Solids* (1969).

Trained manpower and proper equipment are both essential to the fire department if life and property are to be responsibly protected. Yet there is a major factor that works against the firefighter and that is the factor of *time*. Extinguishment methods must be started early. Delay in discovery, delay in reporting the alarm, delay in getting water on the fire, all have been previously cited as factors that contribute to the large-loss fire.

COMMERCIAL FIRE PROTECTION

Early fire detection and extinguishment procedures started immediately have done much to reduce loss of life and property. These two valuable aids, fire detection and fire protection systems, have been incorporated into plant safety by varying methods and means.

Commercial fire protection may be sought through an automatic fixed system that works independently of human power, may require human power to operate, or may be a combination of both.

Standpipe Systems

An important adjunct to the on-site personnel fighting the fire and to the fire department upon arrival are fire standpipe systems. A typical system is shown in Figure 13-1. Fire standpipes are often installed in tall buildings, institutional occupancies, public assembly buildings, and some high-hazard locations.

FIRE DETECTION & PROTECTION SYSTEMS

Figure 13-1 Schematic of typical standpipe system. The low-level fire pump is capable of supplying the necessary pressure to reach the uppermost level of the low-level system with a minimum of 20 psi pressure, and more ideally, 40 or 50 psi. When water is required in the high-level system, the low-level pump discharges its water under pressure to the high-level pump. The high-level pump can then build up the necessary pressure to reach the top floors of the high-level system with adequate required pressure. The fire department must determine which system requires water before hooking up to the siamese connection, or, if doubt exists, it will supply both systems. Auxiliary water can be supplied the system by a gravity tank or pressure tank.

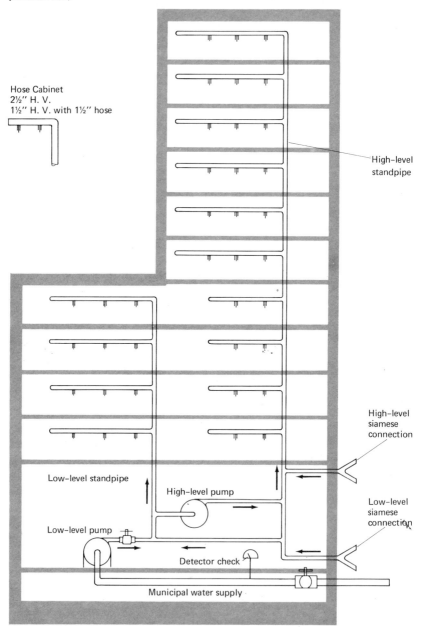

A standpipe is nothing more than a riser water pipe within the structure, of a size determined by the estimated fire flow needed. In general a 4-inch riser is the minimum size for use of 2½-inch hose, with a 6-inch riser required for taller buildings. Standpipes are usually found in stairwells, or, in large floor-area occupancies, spaced over the floor area so that any given location is within 100 feet[1] of the standpipe hose connection.

Usually, a 1½-inch hose line will be connected to the standpipe for first-line defense by the occupants or fire brigade members. The fire department will bring in its own 2½-inch hose to connect to the 2½-inch port on the riser.

Normally, the standpipe system will be a wet-pipe system. In some rare cases, it will be "dry." That is, no water is available through the system until it is supplied by a fire department pumper. The wet-pipe system preferably should be supplied by two independent sources of water. The acceptable sources of water supply are:

1. Public mains where pressure is sufficient

2. Automatic fire pumps with 250 GPM minimum for one riser and 500 GPM where there are more than one riser

3. Manually controlled fire pumps in conjunction with pressure tanks

4. Pressure tanks with a 4,500-gallon minimum

5. Gravity tanks with a 5,000-gallon minimum

6. Manually controlled fire pumps activated from the hose station

The derived pressure from the water supply should be a minimum of 20 psi (pounds per square inch) at the highest outlet in the standpipe and preferably 40 to 50 psi when the water is flowing. Because of the greater pressure that may be obtainable at lower openings in the standpipe system, a firefighter should be stationed at the control valve at the standpipe outlet in use to control the pressure being delivered by regulating the valve.

A supplementary supply of water to the standpipe should always be furnished by the fire department. Fire department pumpers can

[1]This distance varies depending on local codes. Many municipalities specify 75 feet.

supply water to the standpipe system by means of a siamese connection at the base of the building. A siamese connection is a two-port check-valve connection. The fire department can augment the system by attaching one or two lines to the siamese ports. At least one siamese connection should be provided for each standpipe system. On large buildings, several will be placed at various locations for easy access by the fire department.

As previously mentioned, fire department 2½-inch hose will be carried into the fire building and connected to the proper standpipe outlet inside.[2] Simultaneously, the pump operator should be connecting a line to the siamese connection to pump water into the system where needed.

Sprinkler Systems

An automatic sprinkler system is similar in construction and supply to the standpipe. By means of a series of sprinkler heads, it is designed to automatically place water on a fire so as to extinguish the fire or to hold it in check until further aid can arrive. The sprinkler head is fed by a series of pipe lines graduated in size.

Automatic sprinkler systems are credited with being the best counterattack against fire and have permitted structural conditions and occupancies that would not be feasible or safe without their use. In fact, sprinkler systems have an admirable record of efficiency: in 96 percent of the fires involving fires, the sprinkler has extinguished or controlled the fire. Most failures of sprinkler protection are due to human failure rather than system failure.

The sprinkler head is heat-activated, designed to operate at varied temperatures dependent upon the normal temperature of the area to be protected. (See Table 13-1.) Normally, the area covered by one head is 100 square feet. There is an older type of head, in use before 1953, with a smaller deflector head that covers only 80 square feet. This head may still be found on the older installations.

There is a possibility of water damage from excess water after fire extinguishment. When the system is activated, an alarm bell rings to alert the occupants to the probable fire. This warning also aids, by bringing a fast response, in getting the water supply shut down when not needed.

---

[2]As noted in Chapter 11, firefighters using elevators should stop the elevator one or more floors below the fire floor.

Table 13-1 Temperature Ratings of Sprinklers

MAXIMUM ROOM TEMPERATURE AT CEILING	OPERATING TEMPERATURE		COLOR
	NONSOLDER TYPE	SOLDER TYPE	
100°F	135°–150°	155°–165°	Plain bronze
150°F	175°	212°	White
225°F	250°	286°	Blue
300°F	325°	360°	Red
375°F	400°*	415°†	Green
475°F	500°*		Orange

*Special: available only in quartzoid type.
†Special: available only in chemical type.

Table 13-2 Discharge Table of Standard Sprinklers

FLOWING PRESSURE POUNDS (in psi)	DISCHARGE (in gpm)
5	12.80
10	18.10
20	25.80
30	32.80
40	37.20
50	42.20

In small installations, the sprinkler system will be controlled by one main shutoff valve. In larger installations, there will be branch shutoffs controlling various sections or floors of the installation. The fire section can thus be shut off without affecting the balance of the system; sprinkler protection to other portions of the plant can continue.

The standard sprinkler head has a ½-inch discharge orifice. (See Table 13-2 for the pressures and discharges of standard sprinklers.) Minimum pressure at the individual orifice is 7 to 8 psi, with 15 psi being more desirable. At this pressure, the delivery rate of water in gallons per minute should be 22 gpm. The formula used to derive this is

$$\tfrac{1}{2} \text{ psi} + 15 = \text{gpm flow}$$

If too many heads are opened because of the extent of fire, it will be difficult or impossible to obtain adequate flows from the opened orifices. The design of the piping is such that it normally cannot effectively operate with an excessive number of open heads.

At any rate, any good flow of water for a normal situation depends on a reliable source of supply. (Figure 13-2 shows a system with five water sources.)

The acceptable water sources are:

1. Municipal water system mains

2. Gravity tanks, with a minimum 5,000-gallon capacity with bottom of tank not less than 35 feet above the highest sprinkler head

3. Pressure tanks, with a minimum capacity of 4,500 gallons. Filling the tanks two-thirds with water, and having an air pressure of 75 psi, gives a supply of 3,000 gallons

4. Fire pumps, with their capacity, type, and source of water determined by the installation

5. Fire department siamese connection

Sprinkler systems come in various types that, in general, fall within one of four classifications.

WET-PIPE SYSTEM

This system always contains water under pressure in the piping; it will flow when a head is opened.

DRY-PIPE SYSTEM

Where there is a danger of freezing, this system is used. The piping contains air under pressure in place of water. If a head is fused, the air pressure is released, permitting a remote dry-pipe valve to open. This open valve permits water to enter the system and flow to the opened heads. There is, of course, a delay in water on the fire compared with a wet-pipe system, but there is the protection against burst pipes from freezing.

PREACTION SYSTEM

A preaction system is designed to protect installations that would suffer from water damage when water is released without fire. A safety feature is added. Water will not be released through the open head or through faulty piping until a heat-detecting device is activated that opens a preaction valve controlling the flow of water.

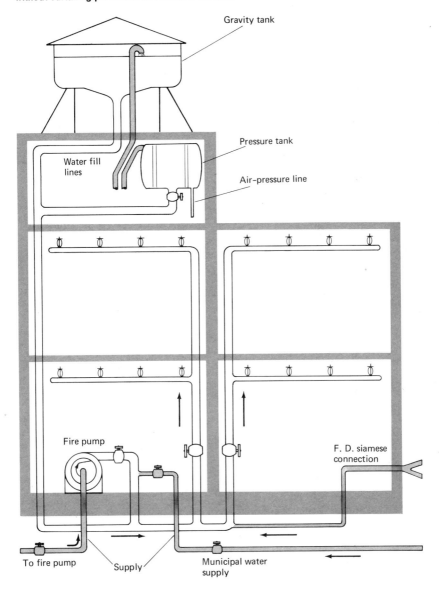

Gravity tank

Pressure tank

Water fill lines

Air-pressure line

Fire pump

F. D. siamese connection

To fire pump

Supply

Municipal water supply

DELUGE SYSTEM

A deluge system has all heads open at all times. The water is controlled by a deluge valve, similar to the preaction valve, that is activated by a heat-detection device. When this valve is activated, there will be a "deluge" of water delivered to the occupancy being protected. Normally, this is an occupancy of greater-than-average hazard.

These, in essence, are the main classifications of sprinkler systems. Any given installation can have a combination of several types or a modification of a given type, depending on the occupancy to be protected.

As in the case of the standpipe, the sprinkler system also provides for a means by which the water supply can be boosted by a fire department pumper. Authorities disagree as to what priority should be given to the fire department siamese connection. All concur, however, that if the first line dropped at the fire ground is not placed into the siamese connection to further supply the system, then the second line should be.

It is imperative that the fire department know the sprinklered buildings in its area, that its firefighters know where the siamese connection is located and also where the main and branch shutoffs are located. To prevent more water damage than fire damage, the firefighter must close the shutoff to stop the flow of water. In those cases where the fire is still in progress, caution must be used in shutting down sprinklers. The fire must be under control by other means first.

Special Fire Protection Systems

Any of the extinguishing agents previously mentioned can also be used in a fixed, automatic extinguishing system where the peculiar characteristics of the specific extinguishing agent can be used best in protecting against the hazard to be covered.

These systems supply extinguishment potential that normally brings the fire under control before the fire department arrives. They cut the time factor to the bare minimum between fire inception and "water on the fire."

WATER SPRAY

A water-spray system is similar to a sprinkler system in that a series of piping delivers water through an orifice over and around a particular hazard. The system may be manual and will have to be turned on when needed, or it may be fully automatic. This system is designed to protect flammable liquid and gas storage, drum-storage areas, large electrical transformers, and some explosive manufacturing processes.

Ordinarily, total extinguishment may not be sought, but rather, controlled burning by keeping containers cooled against expansion and explosion. There would be a great danger of a further disastrous explosion if the leaking gas or liquid were extinguished and then again ignited by another source of ignition. Under the controlled burn, it may be possible to shut off the supply, or burning can continue until the liquid or gas burns itself out.

CARBON DIOXIDE SYSTEM

This type of system, automatically activated, has many advantages. It is comparatively damage-free as there is no liquid or solid residue. There may be prompt resumption of work in those fires where there has been no deep-seated burn. Because of the low cooling effects, it may be necessary, in the case of deep burning, to keep the room closed up and the inert atmosphere maintained until all danger is past. The CO_2 is normally released in a sealed-off area so that the inert atmosphere remains.

Because of the negligible damage of the extinguishant, it is used for high-value rooms—vaults, computers, record rooms, and other such areas.

When the system is activated, the CO_2 "snow" soon retards vision and there is danger in the oxygen deficiency of the atmosphere. For this reason, a warning signal, usually 30 seconds, sounds before the system triggers itself so that the room can be evacuated and closed.

DRY CHEMICAL

This is a rapid extinguishing system for surface fires and is in use over flammable liquids, such as dip tanks, liquid storage, or liquid-spill areas. Being nonconductive, it is useful for electrical equipment but can damage delicate installations such as switchboards and computers. The expellant used to discharge the dry chemical is normally nitrogen gas.

A growing use of a dry chemical extinguishing system on smaller installations is its use in restaurants or cooking facilities. It is used in range hoods, in exhaust duct systems, and over deep-fat fryers. In this type of installation, care must be taken to check for fire that may have extended past the installation or communicated to adjoining framing areas of the ductwork.

FOAM

A fixed foam installation is the ideal installation for outdoor storage tanks. It is also used over localized hazards within the building proper. The system is designed to cover the entire floor area with foam. The foam used may be mechanical or chemical.

The foam may be piped from a central foam house to the tank outlets, or the foam-producing unit may hook up to the tank discharge piping when the fire incident occurs. (See Figure 13-3.) Variations may use water in conjunction with the foam or be a "light water" system, discussed in Chapter 8 on extinguishing agents.

HIGH-EXPANSION FOAM

This is a total flooding system using the high-expansion "soap bubble" foam where a large area must be protected.

Figure 13-3 The formation of foam in a fixed installation. Foam can be introduced either above or below the surface of the liquid being protected.

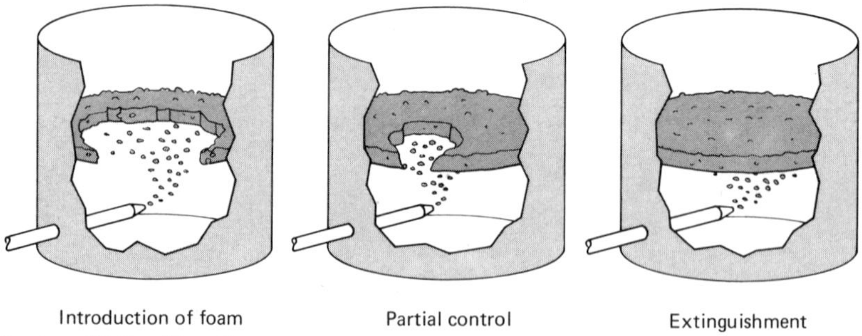

Formation of foam blanket during subsurface injection

Introduction of foam   Partial control   Extinguishment

Courtesy of National Foam Company.

Aircraft hangars (see Figure 13-4), record storage, computer rooms, and warehouses are some protected installations. The damage to the flooded hazard is minimal in this system also.

HALOGENATED EXTINGUISHING SYSTEMS

As previously explained, the halogens inhibit combustion. They can be used in total flooding or local applications and may be installed in auto racing cars and aircraft engine nacelles. They are being tested for use in the oxygen-enriched atmosphere, a particularly hazardous environment.

Figure 13-4 High-expansion foam fixed system protecting flight-test hangars of the L1011 aircraft at Lockheed, Palmdale, California.

Courtesy of Kidde, Inc.

Detection and Alarm Systems

Other means of great value in diminishing the time lag between incipiency and start of extinguishment are various types of fire-detection and warning systems. The systems have different functions and can be classified as to use and purpose.

1.   A local system is one designed to alert the building occupants only. It may designate the area of the incident within the building proper, but it will not summon outside help. It is necessary that some building employee or the watchman-guard be delegated the responsibility of calling the fire department. Too often, another time lag occurs when this person either checks the incident before calling for help or neglects to do so altogether because he or she considers the incident too minor or becomes involved in company fire-fighting tactics.

2.   A proprietary system is designed for a large building or plant facility. Here the alarm signal will go to a central location, usually a part of the plant facility. Plant safety personnel then take appropriate action. Again, no alarm is given to the fire department until the personnel at the central location take that action.

3.   The central station system is under the control and guidance of an outside agency specializing in watch services. The alarm, in addition to alerting building occupants, will register in the central watch station. The watch service will then notify the fire department. This watch station may be some distance from the incident location and it will also be supervising many other building locations within the general area.

4.   The auxiliary system ties the local plant system in with a fire department system. It will alert the fire department at the same time that it gives the local alert. It is the ideal system for schools, hospitals, nursing homes, and public assembly buildings where a life hazard exists and where there might be a delay or failure in notifying the fire department.

5.   The remote station system, usually leased, alerts either a watch service agency or the fire department.

Some warning systems have manual-pull stations that must be manually activated by the person who discovers the fire. In some systems, the manual-pull station is tied in with automatic systems of detection. Other systems may be completely automatic.

TYPES OF DETECTORS

Many types of automatic fire detectors have been developed, each type activated by a particular stage of the fire in progress. Some detector systems may in turn activate fire control equipment that has been installed in conjunction with the system.

The sensitive-to-heat detector was the first type to prove successful. This detector is a thermostatic type that operates either on a rate of rise in temperature or on the attainment of a fixed temperature. Some sensitive-to-heat detectors will reset themselves when the temperature drops, others need to be replaced. This at present is the most widely used, is easy to install, and is relatively inexpensive.

Detectors that operate on readings of infrared and ultraviolet rays detect fire when it is in a flame stage. The flame type of detector is most often used for specific installations and is usually not found in general use.

Photoelectric-beam or spot-type detectors detect fire in the smoke or smouldering stage. The smoke obscures the beam ray, and at a given buildup point, the alarm goes off. The use of this type is becoming more prevalent and can do much not only for life safety but also for lessening smoke and fire damage through early detection.

An ionization detector is the latest development and of the greatest value. Using the ionization principle, a fire can be detected in its incipient stage before smoke, heat, and flame have developed. This type has an ionization chamber that holds minute amounts of an alpha ray–emitting substance. Air between plates in the chamber becomes ionized—electrically conductive—from emission of the alpha ray. Combustion particles increase resistance to the ionization current flame and can trigger an alarm. (See Figure 13-5.)

The earlier detection of the fire potential in the incipient stage by the ionization type of detector not only allows time for safe evacuation of building occupants, but gives the fire department more time to get water on the fire before heat, smoke, and flame have built up to flashover conditions.

Standards for all these systems have been established by the NFPA to guide and govern operation and installation. They must include a means to tell whether a system is operable and a reliable primary power source, and they must be regularly tested and inspected.

Figure 13-5   A view of an ionization type of detector system and a cross section of the inner mechanism.

Courtesy of Pyrotronics, Inc.

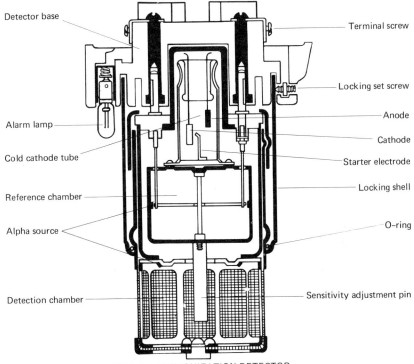

Detector base

Terminal screw

Locking set screw

Anode

Cathode

Alarm lamp

Cold cathode tube

Starter electrode

Locking shell

Reference chamber

Alpha source

O-ring

Detection chamber

Sensitivity adjustment pin

PYR-A-LARM IONIZATION DETECTOR
Model DIS-5B

Watchman Service

Guard or watchman service is employed in many industries, plants, and factories. One function of the guard service is to protect the property against the threat of fire both when occupied and when closed to normal operation. This function includes immediate reporting of any fire incident.

The guard or watchman may be an employee of the building management or of an outside company hired to supply the service. To ensure plant safety, it is necessary that the entire building area be patrolled. These patrol tours are usually supervised—supervised to the extent that the guard may be required to record patrol stations visited by means of a keyed watch clock. Each station has a key that activates a recording device in the clock to indicate the time of station visits.

Under another method of supervision, the guard may be required to activate patrol stations that record a signal in a central location within the plant (if it is large enough), or to a central watch-service office outside. When signals are not received on time, investigation can be immediately made to determine the reason for the omission. The guard must be mentally capable of using good judgment and trained in procedures to be followed in the event of emergency. Needless to say, he or she must have the necessary communication equipment to alert the central office or the proper agency, such as the fire department, when the emergency arises.

Plant Fire Brigades

Many large plants, hazardous occupations, institutional complexes, and public assembly occupancies will establish their own fire brigade. Its members are trained in the use of on-site fire equipment and can do much to control the fire pending the arrival of the fire department. Large complexes may have their own water supply and yard hydrants, hose-cart stations, and, quite possibly, a pumper unit of their own.

The small brigade will be made up of employees who perform another job task for the company when not engaged in fire safety measures within the plant or in training exercises. The larger complex may very well have a full-time fire brigade composed of members whose job role is solely that of firefighter.

Portable Fire Extinguishers

An important means of cutting time to "water on the fire" is by the use of extinguishing agents kept on the fire-ground scene before the arrival of the local fire department. This use can be initiated by plant fire brigades or personnel trained in the use of the equipment.

Extinguishing agents of accepted standard are in practical use in varied ways. Most familiar to the average individual is the use of portable extinguishers installed in various occupancies as the first line of defense against fire. Fire codes list requirements for type and number deemed necessary for the specific structures and occupancies served. The size of the extinguisher can vary, depending upon the expected use and type of extinguishant. The range can go from a 2½-gallon water pump through a 40-gallon foam extinguisher on wheels to a 750-pound, refrigerated carbon dioxide unit.

In the selection of the proper type and size, consideration is given to the combustible hazard and the expected severity. These factors can be matched with the characteristics and capability of the extinguishing agent.

There is nothing wrong with attempting to control or extinguish a fire with portable fire extinguishers, but it is imperative that the fire department also be notified of the incident immediately. Too often, no thought is given to calling the fire department until it is obvious that the first line of defense has failed or has been used up without getting the fire under control.

HOME FIRE-DETECTION SYSTEMS

In recent years, the need for household fire-detection and warning systems has become more apparent. A review of fire death statistics indicates that the major cause of fire deaths is the dwelling fire. Delayed discovery, victims asleep, a smouldering fire that gives off much toxic smoke and gas—all spell out death.

The need here can be plainly seen—again, early detection to alert the victim. Two major factors—public apathy and initial expense—have slowed installation of detection systems. Many agencies in both the public and private sectors are now concerned with, and working on, this problem. With a strong recommendation coming out of the National Commission on Fire Prevention and Control for the use of household detection systems, a greater use should result. However, because of the millions of households in the United States, many installations will have to be made before an appreciable drop in dwelling-fire deaths will result.

SUMMARY

Time is the primary factor that works against the fire department in its attempt to protect life and property.

If fire is to be controlled with the least amount of loss, the time between fire inception and the initiation of extinguishment methods must be sharply reduced.

One means of cutting this time factor is by earlier detection and reporting of the fire. A watch or guard service is one means of detecting fire earlier. Mechanical means of detection include detection systems that operate on various principles. They can be designed to detect fire in the smoke, heat, or flame stage.

A second means of early fire control is by the use of extinguishing agents by occupants or plant personnel *after the fire department has been notified.*

Hand and portable extinguishers vary in size and employ extinguishing agents particularly suitable for the hazard being protected.

Standpipe systems can be employed both by the building occupants and by the fire department upon arrival. The use of a standpipe system cuts fire department lead-out time considerably and is valuable where lead-outs are virtually impossible.

Time can be further cut by the use of automatic systems, activated at the same time as the detection system or by its own heat-activated device. These systems normally bring control before the arrival of the fire department.

The principal automatic extinguishing system is the water-sprinkler system. But all extinguishing agents can be incorporated into an automatic extinguishing system, the peculiar characteristic of the extinguishing agent being matched with the hazard to be handled.

DISCUSSION TOPICS

1.   List the criteria for an efficient guard service.

2.   Discuss the various types of detection systems.

3.   Why are standpipe systems so beneficial to the fire department and a must in many buildings?

4.   There are several types of sprinkler systems in operation. Discuss the various types.

5.   List types of fixed fire protection systems and their specific uses.

RESEARCH PROJECTS

1.   List some installations of fire detection systems in the local area, specifying the type of detector system used.

2.   Study in depth one type of detector system and report on it.

3.   Visit a large plant employing its own fire brigade. Examine water supply, yard hydrants, hose-cart stations, and other devices, and report on your findings.

4.   List several local installations using standpipes or sprinkler systems. What are the available sources of water for each installation?

5.   List several local installations using a fixed automatic extinguishing system other than a sprinkler system. Identify the type of system used.

6.   Visit one of these local installations and report on it.

FURTHER READINGS

Bahme, Charles W., *Fire Officer's Guide to Extinguishing Systems,* National Fire Protection Association, Boston, 1970.

Casey, James F. (ed.), *Fire Service Hydraulics,* 2d ed., Reuben H. Donnelley Corp., New York, 1970, chaps. 8 and 9.

National Fire Protection Association, *National Fire Codes,* Boston, 1969–1979, vols. 6 and 7.

Tryon, George H. (ed.), *Fire Protection Handbook,* National Fire Protection Association, Boston, 1969, secs. 11, 14, 16, 17, and 18.

FIRE PREVENTION

The fire prevention division, bureau, or section, depending on the department size, is concerned with the prime goal of the fire department — the prevention of fire. The small department may have only one person assigned to the fire prevention section of the department's organizational table. The large department should have a fire prevention bureau composed of many inspectors, some specializing in particular hazards, and the supervising command personnel necessary to supervise functions properly. The bureau head, in turn, is directly responsible to the department head.

The goals of fire prevention are outlined in the statement on the intent and purpose of the fire prevention code, as recommended by the American Insurance Association in its *Fire Prevention Code* (1970 edition), section 1.1.

> It is the intent of this code to prescribe regulations consistent with nationally recognized good practice for the safeguarding to a reasonable degree of life and property from the hazards of fire and explosion arising from the storage, handling and use of hazardous substances, materials and devices, and from conditions hazardous to life or property in the use, or occupancy of buildings or premises.

The task of the fire prevention inspector is to implement and enforce the intent of the fire prevention code. He or she will do this by a systematic scheduling of inspections so that all occupancies are visited periodically. Some occupancies, such as hospitals, schools, and commercial and industrial buildings, may require inspections on a quarterly basis. Other occupancies should be inspected twice a year. The normal dwelling occupancy need be inspected only yearly.

FIRE PREVENTION & INVESTIGATION

14

By inspecting the occupancy, the officer will ensure compliance with code laws and ordinances that govern the occupancy. This may necessitate a written order to the owner (see Figure 14-1) or referral to other municipal departments, such as the building department. Much of the inspection will be concerned with maintenance of fire safety features. The condition of fire protection equipment, aisles and exits, fire doors and enclosures will be checked. Many other fire hazards may be found that fall under the classification of housekeeping. Stock piled too high, rubbish conditions, and careless work habits are some of these possibilities.

Two additional benefits result from the fire inspection. One is the knowledge of the building and occupancy gathered by the fire department for future possible fire use. Another is the goodwill and better relations gained when owners or occupants see the interest the fire department has in their plant safety.

Information obtained in the inspection will be incorporated on an inspector's report, usually made in triplicate. One copy will go to the occupancy, one will remain with the inspector, and the third copy will go on file in the fire prevention office.

Figure 14-1 A model form giving formal notice of fire code or ordinance violation.

Fire Prevention Form

Book No. FIRE DEPARTMENT OF THE CITY OF .... Notice No.

FIRE PREVENTION BUREAU

M ....

No. ....

.... (Date)

Your attention is called to a violation of the fire prevention ordinance on premises No. ....

.... of which you are ....

....

....

....

....

....

You are hereby notified to remedy the conditions as stated above within .... {hours / days} from the date of service of this notice or show cause why you should not be required so to do. If, at the expiration of this time, the same conditions exist and no cause aforesaid be shown, such further action will be taken as the law requires.

By order of the

CHIEF OF FIRE DEPARTMENT

....
Inspector.

This form should be printed in triplicate and in book form, with the original copy delivered to the owner or occupant, one copy remaining in the suspense or working office file and the third copy left in the book and later permanently filed.

Courtesy of the American Insurance Association, Special Interest Bulletin No. 223.

Forms will differ from department to department and also vary for different occupancies. (A form for a mercantile and manufacturing company, for example, is shown in Figure 14-2.) In the main, the form will list structural features, fire hazards, and fire protection features. A sketch of the occupancy showing construction, dimensions, and fire hazards and fire safety features should be included.

The right to enact and enforce ordinances relative to safeguarding life and property is inherent in state laws granting police powers to regulate persons and property for the safety of the public.[1] This overall authority has been granted to the municipality by state charter; in turn, the municipality has assigned those functions relative to fire safety to the fire department and the fire prevention bureau.

Industrial processes, storage and use of flammable liquids and gases, hazardous chemical storage and uses, and the ever-increasing new problems presented by technological advances bring a real challenge to the fire prevention fire inspector. The hazards indicate the need for advanced training and knowledge that may at times be beyond the scope of the average fire department inspector.

In some plants and occupancies, inspectors are aided by inspections conducted by fire protection engineers who have made a study and recommendations at the request of the plant management or as a part of a regular periodic insurance inspection.

When the bureau is large enough, some inspectors can be trained in certain occupancies and thus can become expert in a specialized field within the bureau. For example, they can concentrate on occupancies like public assembly areas, hospitals, schools, and nursing homes; or on the process hazards of gases, liquids, and solids.

The general inspector, then, need be concerned only with the business, mercantile, and light manufacturing occupancies. Some of these occupancies may also be inspected by an insurance inspection inspector, thus supplementing the fire department inspection.

The issuance of licenses and permits for various occupancies and functions relative to life safety also will fall under the jurisdiction of the fire prevention bureau. One example is shown in Figure 14-3.

---

[1] Previous mention has been made of the separate roles of the federal, the state, and the local governments in fire safety.

Figure 14-2   A model form for an inspector's report of a mercantile and manufacturing occupancy. Other occupancies may require forms that carry specific items peculiar to the occupancy.

Fire Prevention Form 6.

_____Fire Department

INSPECTOR'S REPORT
MERCANTILE AND MANUFACTURING

Date................

Location _____

Owner ....

Owner's Address ....

Occupants and purpose for which used ....

1st Floor ....

2nd Floor ....

3rd Floor ....

4th Floor ....

5th Floor ....

Construction .... Stories .... With
Without } basement...............

Exposures: Protected .... Unprotected ....

Roof Construction .... Condition ....

Attic .... Access to .... Location ....

Vertical Openings:

    Light Wells .... Locations ....

    Dumbwaiter shafts, enclosed or open .... Locations ....

    Stairs, enclosed or open .... Locations ....

    Elevator shafts, enclosed or open .... Locations ....

    Condition of elevator pits ....

Interior Fire Protection:

    Automatic sprinklers .... Condition of valves ....

    Siamese connection .... Condition of sprinklers ....

    Material stored at least 18 inches below pipes ....

    Condition of fire extinguishers .... Date of inspection or recharge ......

    Standpipes and hose .... { Valves accessible
    { Condition of hose

    Interior fire alarm .... Date of last test ....

    Other fire detecting or extinguishing appliances:

        Type .... Condition ....

    Fire doors .... Fusible links ....

        Operating condition ....

        If not automatic, are they kept closed ....

Heating System: Kind .... Fuel used ....

    Storage arrangements ....

    Condition of stoves, boilers or furnaces ....

    Mounting ....

    Storage of ashes ....

    Protection above and around furnace and flue pipe ....

    Furnace room enclosed or open ....

(over)

Courtesy of the American Insurance Association, Special Interest Bulletin No. 223.

Light and Power: Voltage used ....
    Location of entrance switch ....
    Fuses examined ....
    Branch fuses of proper size ....
    General condition ....
Trash: Facilities for storage ....
    Frequency of removal ....
Flammable Liquids and Gases: Kind ....
    Location ....
    Amount ....
    Storage arrangements ....
    Permit provided ....
Egress Facilities:
    Exit stairways .... Condition of doors ....
    Open stairways .... Unencumbered ....
    Fire escapes .... Condition ....
    Exit lights .... Visibility ....
Conditions noted in violation of ordinance provisions ....
....
....
....
....

Indicate below floor and elevation sketches showing doors, elevators, stairs, partitions and locations of especially hazardous materials.

Signed ....

Figure 14-3 A model form for a permit for the keeping, storage, use, manufacture, handling, transportation, or other disposition of flammable, combustible, or explosive materials.

Fire Prevention Form 3.

FIRE DEPARTMENT
CITY OF_____

PERMIT

For Keeping, Storage, Use, Manufacture, Handling, Transportation, or other Disposition of Flammable, Combustible, or Explosive Materials, as stated below:

No._____

_____
(Date)

TO WHOM IT MAY CONCERN:

By virtue of the provisions of the Fire Prevention Ordinance of the City of _____, _____

(Name of Concern)

No. _____ Street, _____ conducting a _____

(Business)

having made application in due form, and as the conditions, surroundings, and arrangements are, in my opinion, such that the intent of the Ordinance can be observed, authority is hereby given and this PERMIT is GRANTED for _____

This PERMIT is issued and accepted on condition that all Ordinance provisions now adopted, or that may hereafter be adopted, shall be complied with.

THIS PERMIT IS VALID FOR _____.

This permit does not take the place of any License required by law and is not transferable. Any change in the use or occupancy of premises shall require a new permit.

_____
Chief of Fire Department

THIS PERMIT MUST AT ALL TIMES BE KEPT POSTED ON THE PREMISES MENTIONED ABOVE

Courtesy of the American Insurance Association, Special Interest Bulletin No. 223.

PREPLANNING AND FIRE PREVENTION INSPECTIONS

The fire service and allied interests, both in the public and private sectors, are increasingly concerned with the role played by the fire extinguishment or suppression segment of the fire department in fire prevention measures.

Most fire departments have some form of prefire plan inspection that is followed by the individual fire company within its response area. On these prefire plan inspections, the fire company will check the following:

1. Location and quantity of water supply

2. Construction and occupancy of building

3. Means of egress

4. Possibility of fire spread and exposure hazards

5. Special hazards

6. Auxiliary fire-fighting means, such as sprinklers and other systems

The list is not all-inclusive or complete. In addition, those infractions against good fire prevention practices broadly classified as housekeeping violations, blocked exits, rubbish conditions, and so on, are pointed out, and usually immediate corrections are made. In some departments, code violations outside the realm of housekeeping are also pointed out and cited. In other departments these matters are referred to the fire prevention bureau for necessary action.

The information gathered in a prefire plan inspection certainly aids the fire company in its tactical plan if fire should occur in the occupancy. It can help in aiding in the protection of the lives not only of the occupants, but of the firefighters as well, through the first-hand knowledge of the building environment gained.

A form similar to the regular fire prevention form will also be completed on the prefire plan inspection (Figure 14-4). It will be tailored more toward the use of the initial response company in planning tactical procedures if a fire does occur.

It is now believed that these inspections must be expanded — that more of them must be made and that they must take on more of the aspect of a fire prevention inspection.

There is the growing feeling that the fire company should take on the inspection of all small-unit properties, citing code violations and following through on their compliance. All firefighters thus would need some fire prevention training which would be put to use not only in the company inspection program but also on the fire ground when fire occurs. (Many departments now require newly promoted officers to serve a period in the fire prevention bureau, an experience of inestimable value when they return to fire company duty.) If special problems were anticipated or encountered in the inspection, help from a fire prevention bureau inspector could be sought.

The prefire plan inspection is normally conducted in a relaxed atmosphere, accompanied by a responsible member of the occupancy being inspected. It should generate a feeling of mutual interest in a problem of life and property safety. Suggestions are given and many facts obtained, such as sprinkler shutoffs and gas and electrical control shutoffs. The prefire plan inspection is mutually helpful and usually leads to good public relations.

The fire prevention inspection has the connotation of punitive action that is hard to dispel. It is as hard to convince the occupancy owner who has just been told to spend time and money for a fire safety improvement that it is for his or her own good as it is to convince the two-year-old who has just received a thwack across the buttocks that it also is for his or her own good.

The public is apathetic to fire and is generally not overly concerned with fire safety measures until a disaster involving a large loss of life occurs. At such a time, new legislation usually is proposed to correct the hazard and, because of the catastrophe, is generally accepted by the public. It is imperative that the public become aware of life safety measures before the disaster if loss of life and property is to be cut.

The need for public awareness, then, offers another challenge in the area of fire prevention measures to the local fire company unit.

Figure 14-4 A simple form used in a company preplanning inspection. It can be readily completed by the company office, yet furnishes valuable information for tactical planning.

F. D. 461

BATT. # ______

PLATOON # ______

THIS IS A FIRE PRE-PLANNING INSPECTION.
NOT A FIRE PREVENTION BUREAU INSPECTION!

RICHARD J. DALEY, Mayor ROBERT J. QUINN, Fire Commissioner

FILE #

PREMISES ______ OWNER ______ ADDRESS ______

DATE ______ TIME: ______ A.M. P.M. COMPANY NO. ______ OFFICER ______ INFORMATION FROM ______

AREA ______ HEIGHT ______ CONSTRUCTION ______ OCCUPANCY ______

NUMBER OF UNITS ______ NUMBER OF OCCUPANTS ______

EXIT/S
( ) Obstructed or locked

HEATING SYSTEM
( ) Needs repair
( ) Fuel improperly stored
( ) Rubbish in boiler room

FIRE PROTECTION
( ) No Smoking Signs
( ) No Smoking Rules violated
( ) Extinguishers need Recharging
( ) No Fire Extinguishers
( ) Extinguishers inoperative

MAINTENANCE & ORDER
( ) Rubbish ______
( ) No metal cans
( ) Improper storage of flammable liquids
( ) Improper storage of flammable material
( ) Improper use of flammable liquids
( ) Storage under stairs illegal
( ) Repair plaster

FIRE DOORS
( ) Obstructed
( ) Blocked open
( ) Inoperative

SPRINKLERS
( ) Heads need clng.
( ) Obstructed by stock
( ) Valves closed
( ) System inoperative
( ) No drain signs
( ) No shuft-off signs
( ) Siamese connection obstructed
( ) Siamese connection inoperative

ELECTRICAL
( ) Maintain EXIT Signs
( ) Fire Alarm System inoperative

STANDPIPES
( ) Inoperative
( ) Hose missing
( ) Inaccessible
( ) Siamese obstructed
( ) Siamese connection inoperative
( ) Couplings inoperative

FIRE ESCAPES
( ) Obstructed
( ) Need repairs
( ) Need painting

Approved ______ Chief of Battalion

Approved ______ Division Fire Marshal

REMARKS: ______

OTHER HAZARDS: ______

SKETCH AND ADDITIONAL INFORMATION TO BE MADE ON OTHER SIDE!

DON'T HESITATE -- IN CASE OF FIRE CALL FI 7 - 1313 DISPOSITION ______

Courtesy of the Chicago Fire Department.

It has been stated that an intelligent inspection of a complicated plant occupancy is the responsibility of a fire protection engineer rather than of a fire inspector. There are indications that even an intelligent inspection of specified occupancies cannot be made unless the fire prevention inspector has had the specialized training required for that occupancy.

But in the area of the dwelling fire and the family, there is a wide need for education that falls within the range of the local fire company firefighter. This is the area where so many lives are needlessly lost. No complex fire safety measures are needed, but a presentation of common-sense safety fire precautions to the public.

This education has been partially begun by a voluntary home inspection program — a program that needs expanding. In those dwellings that have permitted home inspections, hazards have been pointed out. Self-inspection blanks and home and personal safety folders have been given the families. Recognition of hazards and preventive measures is then taught. A simple home safety program similar to the following has been emphasized:

1. Don't smoke in bed.

2. Close all interior doors at night to prevent passage of smoke, heat, and gases.

3. Make it routine, before retiring, to check to see that doors are locked, lights off, and so on. Check fire safety features as well: Are the stove burners off, the hearth fire out, electrical appliances turned off, and so forth.

4. And, most importantly, develop an emergency escape plan to be followed from each room in the event of fire.

The program of safety education can be expanded outside the home as well. Civic groups and organizations may be made more fire-conscious through efforts of the local fire station. The fire company can present talks and demonstrations to civic groups. It can also bring the public into the fire station for planned activities that promote public awareness of fire safety. The local fire company can aid in school fire-safety measures at the lower, intermediate, and upper grade levels by supervising and participating in fire-safety educational programs.

The means and methods can best be determined by the individual community fire department. This program, coupled with state and federal programs toward the same end, should almost certainly bring a decrease in the loss of life and property.

FIRE INVESTIGATION AND THE ARSON FIRE

There is yet another role that the fire prevention bureau or section fulfills — the role of fire investigation.

The bureau of fire prevention will normally investigate all fires in the municipality that have caused large losses, that have resulted in death or injury, or that are of a suspicious nature.

It is important to try to pinpoint the cause of the fire. An accurate indication of the cause and damage of fire can lead to better statistics and point up the areas in need of greater surveillance. This is especially important in those fires where injury or loss of life has resulted.

Often the investigation is routine and the information is readily apparent at the fire scene or obtainable from occupants or witnesses.

The incendiary fire is another matter. Its solution is much more difficult, and also brings the police department into the picture.

A successful solution to the crime of arson usually requires the joint participation of both the fire and the police departments. Usually, the local fire department, or the state fire marshal, will be the responsible agent for the initial determination of possible arson, fire causes, and collection of evidence. The criminal investigation, collection of further evidence, preevaluation of evidence, and prosecution will then become the responsibility of the police department. Some fire departments (for example, the Los Angeles Fire Department), however, investigate, collect evidence, and prosecute also. In some cases, the state fire marshal is charged with the responsibility of complete investigation, turning only the prosecution over to the state's attorney.

Arson is recognized as one of the most difficult crimes to prove and prosecute. In most criminal actions, it is easy to determine that a crime has been committed. The police can then proceed to a determination of who committed the crime and why. In cases of suspected arson, the first determination, before proceeding to the solution, must be that arson is involved. This determination is often hampered by a lack of evidence, most of it having been consumed in the fire. Further, with some exceptions (noted later), most arson cases have been perpetrated by one-shot criminals who are most difficult to trace down and apprehend.

Arson is a costly crime for some of the same reasons. Salvage and recovery are practically nil. The ashes of the fire have little salvage value.

As in all criminal investigations, the corpus delicti must be established. In brief, this entails:

1. Proof that fire did occur

2. Description of the fire building

3. Indication of fire by criminal design, by either positive o circumstantial evidence

4. Elimination of all accidental sources

5. Connection of the fire to an individual or individuals

Arson has long been categorized as a crime and, as such, punishable. In England in the eleventh century, the arsonist was sentenced to death. At the time of Henry II, arson was punishable by the loss of a hand or foot and banishment from the country. Some recent state laws have also appeared overly severe; life-imprisonment sentences were common. These laws made juries unwilling to convict the person who saw his family in dire straits and used arson as a means to obtain some ready insurance money. To this end, and to broaden the definition of arson as defined by Blackstone as "the malicious and willful burning of house or outhouse of another," the National Fire Protection Association formulated the Model Arson Law.

Model Arson Law

In brief, the Model Arson Law is as follows:

1. Arson, first degree: To burn, aid in burning, or cause to burn the dwelling or buildings attached, whether of himself or of another. Sentence: not less than two nor more than twenty years' imprisonment.

2. Arson, second degree: To burn, aid in burning, or cause to burn buildings other than dwellings, whether of himself or of another. Sentence: not less than one nor more than ten years' imprisonment.

3. Arson, third degree: To burn, aid in burning, or cause to burn personal property other than buildings valued at more than $25 and belonging to another. Sentence: not less than one year nor more than three years.

4. Arson, fourth degree: an attempt to aid in burning, or to cause or attempt to burn, properties as stated in the foregoing classifications. Sentence: not less than one nor more than two years or a fine not to exceed $1,000.

Most states have now incorporated this Model Arson Law, or modified versions of it, into their codes.

Mention has been made of the difficulty of determining whether arson has taken place. The factors involved may be recapitulated thus:

1. The possibility of "clues" being destroyed in the fire itself.

2. The destruction of clues by the fire department in extinguishing and overhauling the fire.

3. The practice of many fire departments of assigning a natural cause of ignition, such as "faulty electrical" or "cigarettes," to the fire of possibly undetermined origin.

4. The lack of training of fire-fighting personnel in detection and conservation of evidence.

To these factors may be added the multiplicity of motives and the heterogeneous character of the arsonist.

Classification of Arson Fires

To aid in the determination of whether arson has taken place, a broad classification has been made of an arsonist's possible motives.

1. Arson for profit — the "gain" fire

2. Arson for revenge — the role of hate, spite, and anger

3. Arson — by the juvenile, the fire bug, or the pyromaniac

4. Arson as a coverup to retard the detection of other crimes

The following pages will examine the individual classification and the emerging character. A clear-cut delineation cannot be made and there are overlaps. An examination can be made only in the broader sense. The arsonist who sets a fire for personal gain by burning out a competitor certainly may also fall under the classification of the pyromaniac.

THE GAIN FIRE — ARSON FOR PROFIT

It is an interesting fact that in times of economic recession, the percentage of fires in the small business, store, or office increases. It appears that the business owner who has a large stock on hand that is not moving, who sees need for expensive retooling to update the product, or who has an old building and business on land that is worth more as vacant property is particularly susceptible to committing arson for personal gain.

Contributing to this classification is the practice of many insurance agents to overinsure accounts for their own personal gain through the commission involved. In many cases, no inspection is made to determine the value of the contents insured.

This type of arson-for-profit fire may also be the work of:

1. The competitor who seeks to eliminate his competition

2. The disgruntled partner who sees a way of dissolving the partnership

3. The seeker for "new business" or "for employment"

In this third class can be found, for example, the carpenter who will submit a bid for the fire repairs necessitated, the watchman or guard who may be hired so that fire will not occur again, the insurance agent who can write a bigger policy with better coverage. All these persons have a profit motive in mind — at the expense of the insurance carrier and the innocent insured.

Running through this classification of arsonists, and occurring in other classifications as well, is another type of arsonist who is motivated for profit — the professional arsonist. This person usually is a member of an organized arson ring — a part of organized crime in America. The motivation of the professional arsonist can be intimidation, or pure hate or revenge.

Like other occupations, arson has certain job hazards. One "torch" became overly zealous in his labors and torched himself in the basement of a restaurant in Chicago; he was found badly burned on the scene by the first arriving fire company and died later at the hospital.

ARSON FOR REVENGE

In its broadest sense, the revenge fire is motivated not by desire for profit, direct or indirect, but by anger, hate, and spite. Strikes and racial and religious overtones may be added. It is within this classification that the sociological factors of the culture and environment can be seen interacting and, in many cases, directing the anger, hate, and spite. Of course the psychological makeup of the individual also plays its part.

Anger is something that usually builds up quite suddenly and quite often is equally as quickly dissipated. The "anger" fire is usually one that is set immediately while anger is at its peak and the perpetrator is usually known by the victim. In fact, often the victim has been told by the angry arsonist that he will burn him out.

Usually it is the jealous lover or spouse who makes the threat, goes to the nearest gasoline station for a can of gas, pours it on the front or rear entry, and applies the match. Unfortunately, he or she occasionally does the job too well; it is generally done at night, and someone, trapped inside, dies. Usually this is a fire of the financially deprived — a product of the social milieu wherein today takes precedence over tomorrow, wherein future retribution and justice may not be realized. A wrong, fancied or actual, must be righted immediately, with the sufferer acting as judge, jury, and executioner. But again, economic lines may be crossed, as exemplified by the middle-class businessman who torched the kitchen of his home in anger against the wife who had threatened to leave him.

When anger has cooled but not totally dissipated, it may be replaced by spite. Unlike anger and the momentary passion it engenders, spite lies dormant under the surface, ready to strike when the opportunity is presented. The person who, when walking a dog in the alley, sees the loaded trash can adjacent to the garage of a disliked neighbor is not really an arsonist at heart if he drops his cigar butt in the trash can. But if the trash can smoulders and then sets the garage on fire, he believes that the fire is really the garage owner's fault. The trash can never should have been there — besides, its owner has asked for it. Spite and the opportunity to "get even" have made arsonists out of many otherwise stable citizens. The intent is a punitive one of inconvenience or small monetary loss rather than large retributive action. It is these fires that are usually marked "carelessness" by the fire department and do not enter the statistics on arson fires unless observed in their inception.

Anger that is not dissipated can spread like a cancerous growth into hate and a desire for revenge. The arson fire that stems from hate and revenge carries implicitly with it the desire to kill, maim, or damage. Like cancer, this too can remain latent for many years, then be manifested. If the opportunity does not present itself, the opportunity will be created. The arsonist in this classification must also be categorized as a psychopathic or neurotic individual.

All the examples in this revenge category have been individual actions. Within the pale of revenge arson also falls the group action of labor strikes or racial and religious differences. The history of our country chronicles many mob actions of violence, including fire, in conflicts between labor and management and in religious and racial disagreements.

ARSON BY THE JUVENILE AND THE PYROMANIAC

From a study of arson fires by the individual or group with a "cause," whether that cause be based on hate, anger, envy, or spite, a progression can be made to a study of the arson fire by the mentally afflicted individual — the pyromaniac.

The pyromaniac has been defined as the maniac whose delusions center on fire, the person with a compulsion to start destructive fires. In the broad classifications that follow, this definition is too narrow and confining. The term must also be used to enfold the juvenile seeking tender loving care and the adult or adolescent maliciously seeking excitement or gratification, but not necessarily compulsive or psychotic in character.

Yet, even within an expanded definition of the term pyro, there are similar themes and patterns: There is no rational or rationalized motive. There is no profit motive, no attempt at crime concealment, no racketeering, no sabotage or revenge motive. The deed is usually planned for, and carried out in, an area easily accessible to the general public. If inside, it will be in a school, library, restroom, or some other gathering place, or in the home of the pyro. If outside, it will be in an alley, on a back porch, in a vacant building or building under construction, in a parked car — all easily accessible to anyone but yet with some "cover" from prying eyes. There is usually no elaborate plan. A few matches, some paper, sometimes a flammable liquid. In many cases, the fire will not communicate to other combustibles but will burn itself out. The pyromaniac will usually be in the crowd of spectators at the fire scene if the fire attempt does succeed, or perhaps even actively aiding the firefighters. Quite frequently the fire is a substitutive sexual expression, particularly in the case of the more narrowly defined pyromaniac.

Pyromania knows no age limits. It is this classification in which the juvenile falls, many fires being set by the preadolescent boy or girl. In the main, the cause is found to be the need for attention. The younger child who lacks a stable family life, who suffers the need of nurture and empathy, may seek to be the center of interest caused by a fire in his or her home. The child may lash out in a hysterical reaction to the environment, and in this reaction, may attempt to find the love and care he or she so sorely needs.

The adolescent pyromaniac may also be seeking needed attention, or the act can be an act of rebellion against authority. At this time of physical and emotional change, the fire may also serve as a vicarious sex thrill.

The adolescent may start his or her career as a pyromaniac by turning in false alarms of fire. As the thrill of the false alarm weakens, he or she turns to starting the small fire. If not apprehended, the adolescent may become a pathological firesetter in need of psychiatric care.

A class of male pyromaniac is the "vanity" firesetter. Here the male becomes the "hero" who warns the occupants of the fire building once the fire has started. He is the "would-be firefighter" who aids the firefighters in the extinguishment of the fire. His ego demands bigger and bigger fires with possible loss of life until he is apprehended. One of these vanity firebugs frequented a fire station, assisting in the station chores of housecleaning and hose changing. It was only after a long series of back-porch fires that his daily visits and after-dark departures with the day-old papers under his arm were related to the chain of fire incidents.

The vanity firebug is one of the types of confirmed compulsive pyromaniacs who will start a fire with no concern for personal safety nor for the safety of others.

The classification of abnormal psychological behavior that is manifested by pyromania has many bedfellows. The epileptic, following a major attack, may suffer amnesia, become irritable, troublesome, and, on occasion, a pyromaniac. A psychothemia group of psychoneurotics under emotional tension, with obsessions, fears, and doubts, can experience the compulsion to set a fire. The mentally handicapped may become addicted to pyromania. The alcoholic or drug addict who is normally rational, when drunk or drug-influenced, may be a pyromaniac.

Within the sexual-substitutive group can be found the fetishist who has replaced the urge for female clothing with an urge for fire. The exhibitionist, the suppressed homosexual, the peeping tom, all may use fire as a substitute for the substitute sexual act that is their standard trade.

ARSON AS A COVERUP

Through past ages, arson appears to have been the classic device used to destroy evidence of other crime. The murderer has often attempted to destroy the corpus delicti by arson. If total destruction of evidence of the murder cannot be achieved, he or she can at least hope to hinder or prevent identification of the victim.

The embezzler may see arson as the only means of destroying the records that would reveal the money shortage. Robbers may attempt to conceal their entry and larceny by torching the building. The forced entry and burglary followed by fire, however, usually are the act of an adolescent in rebellion against authority or some segment of society. (Examples are the frequent acts of vandalism plus fire in a school.)

In many instances, arson as a coverup for other crime only brings attention to the original crime. It has been stated that arson is particularly hard to solve because of the destruction of evidence. Yet, in many instances, the evidence of the other crime is not totally destroyed and the fire points up the fact that a crime has been committed. Quite possibly, when arson follows pilferage and embezzlement, the fire aims a finger toward those responsible. For that reason, it appears that an attempt to cover up one crime with another, arson, is usually made only by the novice in crime. The experienced professional criminal will stake his or her chances on getting away with one crime rather than risk exposure to a second crime, and a crime in which he or she is not a professional.

Thus it can be seen that the investigation of an incendiary fire, the arson fire, poses a definite challenge to the fire investigator. It, too, requires special training and an understanding of human psychology.

SUMMARY

One of the main functions of the fire department is to establish and operate an active fire prevention bureau and to enact a fire prevention code.

To ensure the fire safety of the municipality, it is necessary to maintain a continuing inspection program by specially trained personnel.

Because of the varied and technical hazards of certain occupancies, it is desirable to train inspectors for a particular occupancy where feasible. Examples that require special training are hospitals, large public assembly buildings, and highly involved industrial complexes.

The inspection will check for:

1. Compliance with code laws and ordinances governing the occupancy

2. Condition of fire protection equipment, exits, fire doors, and enclosures

3. Housekeeping conditions that affect fire safety and work habits of the employees that also affect fire safety

The local fire company also is engaged in fire prevention inspection work. Inspections of small-unit properties and dwellings can be undertaken by the local fire company. Prefire planning inspections should be made to determine individual occupancy conditions that may affect tactical fire ground planning. The local fire company can do much in creating better public relations and in educating the general public in fire safety measures.

The fire prevention inspector may also be concerned with fire investigation, the determination of fire cause. An accurate determination of fire cause can lead to measures to control the hazards and causes of fire.

The investigation of the arson fire is a particularly difficult task. Evidence may have been destroyed in the fire or in the process of extinguishing it. A motive for, and proof of, arson must be established. Cooperation with the police must be established if the case is to be properly prosecuted.

The motives for arson can cover a wide range. The chief motives are:

1. Arson for profit

2. Arson for revenge, hate, and spite

3. Arson by juveniles, pyromaniacs, and the emotionally disturbed

4. Arson as a coverup for other crime

Arson is a crime that has no age limits, no sex barrier. It is a crime that attracts the amateur as well as the professional. Motivation can range from the lack of mother love to the desire for monetary profit. The perpetrator can range from the emotionally disturbed to the cool, calculating brain of an arson ring. The arsonist may be a lone wolf or a member of an arson ring, but he or she is a criminal.

DISCUSSION TOPICS

1.   What are the intent and purpose of fire prevention and how are they accomplished?

2.   What is the purpose of a prefire planning inspection? How is it best achieved?

3.   Discuss similarities and differences in a fire prevention and a preplanning inspection.

4.   What is the role of fire investigation in fire prevention measures?

5.   The arson fire covers a broad range of both causes and effects. Discuss various classes of arson and the personalities involved.

RESEARCH PROJECTS

1.   Outline a program to promote public awareness of fire safety.

2.   Outline a speech to be given before a civic organization that will call attention to the fire safety problem and public apathy.

3.   List a particular occupancy and the specific fire safety feature that should be checked in this occupancy.

4.   List possible motives for arson and match the motive with an achetypal arsonist.

5.   Cite one or more case histories from the experience of the municipal fire department to match a particular motive for arson.

FURTHER READINGS

American Insurance Association, *Fire Prevention Code,* 1970.

Battle, B., and P. Weston, *Arson, A Handbook of Detection and Investigation,* Arco, New York, 1960.

Huron, B. S., *Elements of Fire Investigation,* Reuben H. Donnelley Corp., New York.

Kennedy, John, *Fire and Arson Investigation,* Rand McNally, Chicago, 1962.

Kirk, Paul, *Fire Investigation,* Wiley, New York, 1965.

National Fire Protection Association, *Burke Procedural and Classification Chart for Arson Investigation,* Boston, 1959.

National Fire Protection Association, NFPA No. 101, *Life Safety Code,* 1967.

CONFLAGRATIONS

A conflagration can best be defined as a fire that communicates to adjoining properties, extending beyond a potential firebreak such as a street or an open area, and involving a goodly number of buildings in fire. The individual plant complex covering a wide area and many buildings is precluded.

Boston in colonial times alone experienced five conflagrations within a little more than a hundred years—from 1653 to 1760. The Chicago fire of 1871, the San Francisco fire following the earthquake of 1906, the Texas City disaster of 1947, Los Angeles and its series of forest and brush fires, the riots of several of the larger cities—all are well-known conflagrations. These noted cases point out the varied underlying causes and the potential yet with us today.

The *Fire Protection Handbook*, published by the National Fire Protection Association, lists factors contributing to the conflagration. They are shown in Table A-1.

A brief examination of this listing reveals that causes existing in the 1926-1967 period of the survey were not apparent in the period from 1901 through 1925. These causes were mainly the transportation or storage of hazardous materials.

Note should also be taken of the role played by an inadequate water-distribution system in the latter period of the survey. The rapid expansion and growth of the country in both industry and population and the resultant fire safety lag are indicated.

As noted, the table does not include combustible construction or contents. Most cities contain large areas of closely built, frame-construction dwellings that offer ready fuel to the spreading fire. However, fire-fighting techniques as we know them today can stop the spreading fire before it becomes a conflagration, unless some factor, probably in combination with several other factors, is also present.

APPENDIX A

Table A-1 Principal Factors Contributing to Conflagrations in the United States and Canada since 1900*

FACTOR	NUMBER OF TIMES CONTRIBUTING		
	1901–1925	1926–1967	1901–1967
1. Wood-shingled roofs	45	24	69
2. Wind velocity in excess of 30 miles per hour, or "high"	22	41	63
3. Inadequate water distribution system	23	32	55
4. Lack of exposure protection	18	29	47
5. Inadequate public protection	23	24	47
6. Unusually hot or dry weather conditions	4	23	27
7. Delay in giving alarm	5	13	18
8. Congestion of hazardous occupancies difficult of access for firefighting	5	13	18
9. Delay in discovery of fire	4	16	20
10. Forest or brush fire entered town	2	10	12
11. Failure of water pumps or breakage of pipes	5	6	11
12. Ineffective firefighting	4	6	10
13. Private fire protection failed or inadequate	1	9	10
14. Fire department at other fires	4	3	7
15. Fire spread through inaccessible spaces under pier or building	2	4	6
16. Severe winter conditions	2	3	5
17. Earthquake, floods, hurricane, etc.	1	3	4
18. Hose couplings or hydrant connections not standard	2	1	3
19. Cotton rags, etc., stored outside of buildings	2	1	3

Factors Contributing to Conflagrations (continued)

FACTOR	NUMBER OF TIMES CONTRIBUTING		
	1901–1925	1926–1967	1901–1967
20. Burning brands from lumberyard	0	2	2
21. Dry vegetation adjacent to buildings	0	2	2
22. Explosion of liquefied natural gas holders	0	1	1
23. Explosion of ammonium nitrate aboard cargo vessel	0	1	1
24. Slow response of fire department	0	1	1
25. Fire alarm failed	1	0	1
26. Explosion of explosives truck	0	1	1
27. Riots preventing or hampering firefighting	0	2	2

*Excluding combustible construction and contents which contribute to every fire.

Source: National Fire Protection Association, *Fire Protection Handbook,* 13th ed.

FIRE REPORT

A good fire report must do much more than establish fire causes, cite an occupancy or industrial hazard, and list materials susceptible to fire. Information in the report must be helpful in locating the trend of fire to specific localities and the hours of greatest incidence. These data can be important in the assignment of the workforce and equipment location. (Refer to the Rand Survey Report for New York City.)

Many more benefits to be derived from a complete fire report can be fully realized only when some form of electronic data processing is used. Many departments now have this facility available to them.

The analysis of the fire report will depend somewhat on the size of the municipality and the information deemed necessary by the municipal heads. However, the basic data to be collected in the good fire report should be similar even if the analysis does not progress beyond number of alarms, working time, cause, and estimate of damage.

The following report, used by the Chicago Fire Department, is an example of a report that lends itself to varied measurements and analysis. Much of the information is given by a check-off or "circling" process. Other information is supplied through the code form that is used in conjunction with the fire report.

This typical report would be more complete if more information were given in a "material ignited" classification.

APPENDIX B

CHICAGO FIRE DEPARTMENT
FIRE REPORT

I. BATTALION NO:_____ STILL CO:_____ DATE:_____ TIME:_____
　　　　　　　　　　　　　　　　　　　　　　　Month　Day　Year　　　0-2400

　　TYPE OF ALARM:　　1. STILL　　2. BOX_____　3. EXTRA ALARM _____
　　　　　　　　　　　　　　　　　　　　　　NUMBER

　　LOCATION
　　OF FIRE:_____
　　　　　　　　NUMBER　　　　　　　STREET　　　　ROOM or APT.　　FLOOR

　　OWNERS NAME _____ ADDRESS_____

II. TYPE OF NON–FIRE: ☐　　1. Water Flow—No Service　　3. False Alarm—Mistaken Citizen
　　　　　　　　　　　　　　　2. False Alarm—Malicious　　4. Fire—No Service　　5. Aircraft Standby
　　　　　　　　　　　　　　If False Alarm or No Service, move to Section XVI

III. TYPE OF FIRE: ☐　　1. Fire in Structure　　2. Fire not in Structure　　3. Fire in Mobile Unit

IV. NON-OCCUPANCY CODE: ☐

　　01　Motor Vehicle　　　　　　　07　Fence　　　　　　　　　　13　Rubbish
　　02　Truck w/Hazardous Material　08　Fire Works　　　　　　　14　Portable Vessel w/Flamm. Gas
　　03　Tank Truck　　　　　　　　09　Grass or Brush　　　　　　15　Railroad Rolling Stock
　　04　Aircraft Fire　　　　　　　10　Lumber (not lumber yards)　16　Roofing Kettle
　　05　Boat or Ship　　　　　　　11　Marina　　　　　　　　　17　Tar Pot
　　06　Bridge　　　　　　　　　12　Pier or Wharf　　　　　　18　Unclassified

V. FIRE IN STRUCTURE: (Use Code Form 163)
　☐　A. Occupancy Code　☐　B. Probable Cause of Ignition　☐　C. Area of Origin Code

VI. FIRE FACTORS:
　☐　A. Building Construction　☐　B. Building Height　☐　C. Building Area
　　　1. Fire Resistive　　　　　　1. One Story　　　　　1. Under 3,000 sq. ft.
　　　2. Heavy Timbers　　　　　　2. 2-3 Stories　　　　2. 3,000-5,000 sq. ft.
　　　3. Non-Combustible　　　　　3. 4-5 Stories　　　　3. 5,000-10,000 sq. ft.
　　　4. Ordinary Joisted　　　　　4. 5-10 Stories　　　4. Over 10,000 sq. ft.
　　　5. Wood Frame　　　　　　　5. Over 10 Stories
　　　6. Combination

VII. BUILDING ALARM SYSTEM: ☐　1. Operated　　2. Not Operated　　3. None on Premise

VIII. EXTENT OF FIRE: ☐　1. Confined to place of origin　3. Extended to adjacent properties
　　　　　　　　　　　　2. Confined to building　　　　4. Extended beyond adjacent properties

IX. FIRE ESCAPES: (Circle all appropriate factors)
　　　　　1. Available & used by Residents　　3. Used by Fire Department
　　　　　2. Available—not used　　　　　　4. None available

X. EXTINGUISHING EQUIPMENT USED: (Circle all appropriate factors)
　　　　　　　　　　　　　　　　　　　　　　09　Hand Pump
　01　Automatic Sprinklers　05　Foam System　　　10　Booster
　02　CO$_2$ System　　　　06　Private Extinguisher　11　1½" Hose
　03　Deluge System　　　07　Standpipes　　　　12　2½ and/or 3" Hose
　04　Dry Chemical System　08　Ansul Extinguisher　13　Heavy Stream Apparatus

XI. SPRINKLERS: (Circle all appropriate factors)
　　1. Controlled Fire　　　3. No Heads Opened　　5. 1-5 Heads　　7. 11-20 Heads
　　2. Did not control Fire　4. None in area involved　6. 6-10 Heads　8. Over 20 Heads

XII. DAMAGE ESTIMATE: (Circle all appropriate factors)
　　TO CONTENTS:　1. None　　2. Small　　3. Moderate　　4. Considerable
　　TO BUILDING:　1. None　　2. Small　　3. Moderate　　4. Considerable

XIII. WEATHER CONDITIONS: (Circle all appropriate factors)
　　　　　　　　　　　　　　　　　　　　　　　5. Temperature over 85°
　1. Clear　　2. Rain　　3. Ice & Snow　　4. Wind a factor　6. Temperature under 15°

XIV. RESCUE—INJURY—FATALITY DATA:
　Number of　　　　　Number of Persons Injured　　　Number of Persons Killed
　Persons Rescued_____　Civilians_____Firefighters_____　Civilians_____Firefighters_____

XV. SALVAGE & ARSON DATA:

　Salvage work performed ☐　Arson Investigation called ☐　Chicago P.D. called ☐

XVI.

　　OFFICER IN COMMAND REPORTING _____
　　　　　　　　　　　　　　　　　　NAME　　　　　RANK　　　　COMPANY

Form F.D. 152

EXCERPTS FROM THE GRADING SCHEDULE FOR MUNICIPAL FIRE PROTECTION

INTRODUCTION

The Grading Schedule is a means of classifying municipalities with reference to their fire defenses and physical conditions. The word "municipality" is used in this Schedule in a broad sense to include cities, towns, villages, or other municipal organizations.

Inherent in the hazards of fire and explosion is the danger to life. Therefore, the elimination or reduction of these hazards and an improvement in fire defenses directly affects safety to life.

From a study of pertinent conditions and performance records extending over many years, certain standards have been developed; these are set forth in the Schedule, and the various features of fire defense in the municipality under consideration are compared with them. For each deviation from these standards, deficiency points are assigned, the number depending upon the importance of the item and the degree of deviation. The ability of the municipality to control hazards by means of appropriate laws and their enforcement is graded in the same way. The total number of deficiency points charged against the municipality determines its relative classification.

Table C-1 shows the features considered, as well as the relative value and maximum number of deficiency points allocated to each.

Additional deficiency points may be assigned for the reasons described below; these points are added to those charged against the features listed in Table C-1.

It is recognized that climatic conditions may increase fire losses by reason of the increased frequency of fires due to the heating hazard, by retarding the response of fire apparatus, by hampering effective firefighting during cold weather and storms, by the increase in combustibility due to hot dry weather, and by the greater probability of fires spreading at time of high winds. Furthermore, occurrences such as earthquakes, tornadoes, hurricanes, cyclones, blizzards, and floods can also adversely affect fire protection. In addition, civil disturbances and harassment of the fire department can reduce the effectiveness of the fire defenses. For such conditions, additional deficiency points are assigned.

APPENDIX C

Table C-1 Relative Values and Maximum Deficiency Points

FEATURE	PERCENT	POINTS
Water supply	39	1,950
Fire department	39	1,950
Fire service communications	9	450
Fire safety control	13	650
	100	5,000

Source: From Grading Schedule for Municipal Fire Protection, Insurance Services Office, 1973. Reprinted by permission.

Where the water supply is considerably better than the fire department, or vice versa, the better feature cannot be utilized to full value. In recognition of this fact an additional deficiency is assigned where the divergence of these features is excessive.

Credit, by which the deficiency points under Fire Service Communications may be reduced, is given for fire alarm boxes in residential sections.

The items considered in grading the various features are listed in Table C-2. Table C-3 shows the relative classes of municipalities and the corresponding range of deficiency points for each class. The remaining pages of the Schedule are devoted to the details for grading the various features.

Under many items, the deficiency to be assigned is determined as a percentage, and this value is translated to a corresponding number of deficiency points. For this purpose a graduated scale is used in which the number of points increases with the percent deficiency but by lesser increments for the smaller percentages. The scale is thus in keeping with the idea that a difference of, say, 10 percent has less actual effect on the protection afforded where conditions are good or moderately good than where they are poor. This scale appears in Table C-2 and is referred to under each item to which it applies; the full scale or a multiple or fraction thereof is used, depending upon the relative importance of the item considered (Table C-4).

When used in this Schedule as applied to buildings, complexes, or districts, (1) "residential" refers to one- to four-family dwellings not exceeding three stories in height, and (2) "commercial" refers to business, industrial, warehouse, institutional, educational, hotel, apartment, and other occupancies not included in (1).

Table C-2 Items Considered in the Schedule

WATER SUPPLY

ITEM NO.
1	Adequacy of supply works
2	Reliability of source of supply
3	Reliability of pumping capacity
4	Reliability of power supply
5	Condition, arrangement, operation, and reliability of system components
6	Adequacy of mains
7	Reliability of mains
8	Installation of mains
9	Arrangement of distribution system
10	Additional factors and conditions relating to supply and distribution
11	Distribution of hydrants
12	Hydrants—size, type, and installation
13	Hydrants—inspection and condition
14	Other conditions adversely affecting adequacy, reliability, or operation of the system

FIRE DEPARTMENT

ITEM NO.
1	Pumpers
2	Ladder trucks
3	Distribution of companies and type of apparatus
4	Pumper capacity
5	Design, maintenance, and condition of apparatus
6	Number of officers
7	Department manning
8	Engine and ladder company unit manning
9	Master and special stream devices
10	Equipment for pumpers and ladder trucks
11	Hose
12	Condition of hose
13	Training
14	Response to alarms
15	Fire operations
16	Special protection
17	Other conditions adversely affecting operations

Items Considered in the Schedule (continued)
FIRE SERVICE COMMUNICATIONS

ITEM NO.

1	Communication center
2	Communication center equipment and current supply
3	Boxes
4	Alarm circuits and alarm facilities including current supply at fire stations
5	Material, construction, condition, and protection of circuits
6	Radio
7	Fire department telephone service
8	Fire alarm operators
9	Conditions adversely affecting use and operation of communication facilities and the handling of alarms
10	Credit for boxes installed in residential districts

FIRE SAFETY CONTROL

ITEM NO.

1	Flammable or compressed gases
2	Flammable or combustible liquids
3	Special hazards
4	Miscellaneous hazards
5	Supplemental fire prevention activities
6	Building laws
7	Electricity
8	Heating and ventilating installations

Source: From Grading Schedule for Municipal Fire Protection, Insurance Services Office, 1973. Reprinted by permission.

Where the requirements are influenced by the heights of buildings expressed in stories, a total height up to 35 feet may be counted as two stories; 36 to 45 feet, three stories; and 46 to 55 feet, four stories.

Where conditions differ widely from those usually found in the average municipality, certain portions of the Schedule may have to be interpreted in a manner consistent with the usual conditions encountered and the extent to which they affect the fire protection problem. Such conditions might be found, for example, in a community experiencing rapid growth or one consisting largely of amusement or seasonal occupancies, or consisting chiefly of buildings of an industrial nature with few residents.

This Schedule supersedes the "Standard Schedule for Grading Cities and Towns of the United States with Reference to Their Fire

Table C-3   Relative Grading of Municipalities in Fire Defenses and Physical Conditions

POINTS OF DEFICIENCY	RELATIVE CLASS OF MUNICIPALITY
0 – 500	First
501 – 1,000	Second
1,001 – 1,500	Third
1,501 – 2,000	Fourth
2,001 – 2,500	Fifth
2,501 – 3,000	Sixth
3,001 – 3,500	Seventh
3,501 – 4,000	Eighth
4,001 – 4,500	Ninth*
More than 4,500	Tenth†

*A ninth-class municipality is one *(a)* receiving 4,001 to 4,500 points of deficiency, or *(b)* receiving less than 4,001 points but having no recognized water supply.

†A tenth-class municipality is one *(a)* receiving more than 4,500 points of deficiency, or *(b)* without a recognized water supply and having a fire department grading over 1,755 points, or *(c)* with a water supply and no fire department, or *(d)* with no fire protection.

Source: From Grading Schedule for Municipal Fire Protection, Insurance Services Office, 1973. Reprinted by permission.

Table C-4   Deficiency Scale

DEFICIENCY POINTS CORRESPONDING TO PERCENT DEFICIENCY

	0%	10%	20%	30%	40%	50%	60%	70%	80%	90%	100%
0%	0	10	25	45	67	90	112	134	156	178	200
1%	1	12	27	47	70	92	114	136	158	180	
2%	2	13	29	50	72	94	116	138	160	182	
3%	3	15	31	52	74	97	119	141	163	185	
4%	4	16	33	54	77	99	121	143	165	187	
5%	5	18	35	57	79	101	123	145	167	189	
6%	6	19	37	59	81	103	125	147	169	191	
7%	7	21	39	61	83	105	127	149	171	194	
8%	8	22	41	63	85	108	130	152	174	196	
9%	9	24	43	65	88	110	132	154	176	198	

POINTS

Source: From Grading Schedule for Municipal Fire Protection, Insurance Services Office, 1973. Reprinted by permission.

Defenses and Physical Conditions" which was first published by the National Board of Fire Underwriters in 1916; subsequent editions were published in 1922, 1930, 1942, and 1956 with amendments in 1963 and 1964.

Modifications have been made to recognize current good practice and modern developments, but these changes should not materially affect previous gradings of municipalities that have progressed with the times. Because of the decentralization that has taken place in most municipalities, more attention is given to the city as a whole with a corresponding decrease in emphasis on the principal business district. The number of grading features has been reduced from 6 to 4, by the combination of Fire Prevention and Building Department into Fire Safety Control and the elimination of Structural Conditions which related only to the principal business district. The number of items has been reduced, and changes in wording and arrangements have been made to make the Schedule more convenient to use.

WATER SUPPLY

An adequate and reliable water supply is an essential part of the fire-fighting facilities of a municipality.

Minimum Recognized Water Supply

In order to be recognized for grading purposes, a water supply shall be capable of delivering at least 250 gpm for a period of 2 hours, or 500 gpm for one hour, for fire protection plus consumption at the maximum daily rate. Any water supply which cannot meet this minimum requirement shall not be graded, and a deficiency of 1,950 points shall be assigned.

Adequacy and Reliability

A water supply is considered to be adequate if it can deliver the required fire flow for the number of hours specified in Table C-5, with consumption at the maximum daily rate; if this delivery is possible under certain emergency or unusual conditions, the water supply is also considered to be reliable.

Table C-5   Required Duration for Fire Flow

REQUIRED FIRE FLOW (in gpm)	REQUIRED DURATION (in hours)
10,000 and greater	10
9,500	9
9,000	9
8,500	8
8,000	8
7,500	7
7,000	7
6,500	6
6,000	6
5,500	5
5,000	5
4,500	4
4,000	4
3,500	3
3,000	3
2,500 and less	2

Required Fire Flow

The required fire flow is the rate of flow needed for fire-fighting purposes to confine a major fire to the buildings within a block or other group complex. The determination of this flow depends upon the size, construction, occupancy, and exposure of buildings within and surrounding the block or group complex; consideration may be given to automatic sprinkler protection.

A required fire flow shall be determined at appropriate locations in each district or section of the municipality. The minimum fire flow requirement is 500 gpm and the maximum for a single fire is 12,000 gpm. Where local conditions indicate that consideration must be given to simultaneous fires, an additional 2,000 to 8,000 gpm will be required. Table C-5 gives the required duration for fire flow.

FIRE DEPARTMENT

General

Fire stations shall be suitable for the purpose and well maintained. The provisions for refueling apparatus at stations and at large fires shall be adequate. The possibility of delays in response due to poor conditions of roads, including inadequate snow plowing and sanding, steep grades, vehicle parking, traffic, harassment, railroad grade crossings, and similar features shall be considered.

Determination of Companies Required

The number of engine and ladder companies shall be sufficient to provide reasonable protection to the municipality.

In determining the number of companies required, consideration shall be given to the distribution needed to provide adequate coverage of the municipality, the overall fire potential, the population and total area served, and any other factors which may have an effect on protection requirements.

Engine and ladder companies shall be located so that the travel distances for the first due and first alarm companies shown in Table C-6 for the respective required fire flows are not exceeded. The maximum number of companies needed for the respective required fire flows shall be located so as to meet the recommended travel distances.

Structural conditions and hazards within the municipality may call for more companies than are determined by the foregoing in order to permit proper handling of multiple alarms.

The distances shown in Table C-6 shall be reduced if a severe life hazard exists; if traffic routing and congestion, topographical features, man-made barriers such as railroad and highway structures, or other local conditions hinder response; or if other circumstances peculiar to the municipality or particular district indicate that such a reduction is needed.

The probability of simultaneous fires, the number and extent of runs, and the need for placing additional companies in service or relocating companies during periods of high frequency of alarms shall be considered. Consideration should be given to providing some protection for all areas during multiple alarms or simultaneous fires which require the response of all or a large portion of the companies regularly in service.

Where protection is furnished to territory outside the corporate limits, additional companies may be required, depending upon the size of the area served, the number and extent of runs to that territory, the amount of service provided, and contractual or other agreements that have made.

Table C-6 Number of Engine and Ladder Companies Needed within Travel Distance of Established Fire Flow

FIRE FLOW	FIRST DUE				FIRST ALARM				MAXIMUM MULTIPLE ALARM			
	ENGINE		LADDER		ENGINE		LADDER		ENGINE		LADDER	
gpm	No.	Mi.	No.	Mi.	No.	Mi.	No.	Mi.	No.	Mi.	No.	Mi.
less than 2,000	1	1½‡	1*	2§	2†	4	1*	2§	2†	4	1*	2§
2,000	1	1½‡	1*	2§	2	2½	1*	2§	2	2½	1*	2§
2,500	1	1½	1*	2	2	2½	1*	2	2	2½	1*	2
3,000	1	1½	1*	2	2	2½	1*	2	3	3	1*	2
3,500	1	1½	1*	2	2	2½	1*	2	3	3	1*	2
4,000	1	1½	1	2	2	2½	1	2	4	3½	1	2
4,500	1	1½	1	2	2	2½	1	2	4	3½	1	2
5,000	1	1	1	1½	2	2	1	1½	5	3½	2	2½
5,500	1	1	1	1½	2	2	1	1½	5	3½	2	2½
6,000	1	1	1	1½	2	2	1	1½	6	4	2	2½
6,500	1	1	1	1½	2	2	1	1½	6	4	2	2½
7,000	1	1	1	1½	2	1½	1	1½	7	4	3	3½
7,500	1	1	1	1½	2	1½	1	1½	8	4½	3	3½
8,000	1	1	1	1½	2	1½	1	1½	9	4½	3	3½
8,500	1	1	1	1½	2	1½	1	1½	9	4½	3	3½
9,000	1	¾	1	1	3	1½	2	2	10	4½	4	4
10,000	1	¾	1	1	3	1½	2	2	12	5	5	4½
11,000	1	¾	1	1	3	1½	2	2	14	5	6	5
12,000	1	¾	1	1	3	1½	2	2	15	5	7	5

*Where there are less than five buildings of a height corresponding to three or more stories, a ladder company may not be needed to provide ladder service.

†Same as first due where only one engine company is required in the municipality.

‡May be increased to 2 miles for residential districts of one- and two-family dwellings, and to 4 miles where such dwellings have an average separation of 100 feet or more.

§May be increased to 3 miles for residential districts of one- and two-family dwellings, and to 4 miles where such dwellings have an average separation of 100 feet or more.

Source: From Grading Schedule for Municipal Fire Protection, Insurance Services Office, 1973. Reprinted by permission.

DEPARTMENT STAFFING

There shall be 6 members on duty for each of the required engine and ladder companies. (See Note B.) One member less need be provided for each hose company considered creditable in lieu of an engine company. Drivers or operators may be needed for special apparatus.

Note A. On-Duty Strength. The total number of members on duty shall be taken as that normal during vacation periods less average details and sick leaves, but not the absolute minimum that may occur on only one or two days a year.

Chief's aides may be included in department strength if they participate in actual fire-fighting operations.

Members on apparatus not credited under Items 1 and 2 that regularly respond to alarms to aid other companies may be included in this item as increasing total department strength. Members on fireboats shall be considered only in Item 16.

Personnel manning ambulances or other units serving the general public may be credited, the amount depending upon the extent to which they are available and are used for response to fire alarms to perform fire duty.

Note B. Call and Volunteer Members. In departments having call or volunteer members, four call or volunteer members, on the basis of the average number responding to alarms, may be considered as equivalent to one full-paid member on duty, but the number of such equivalent full-paid members shall not exceed one-third the required strength of existing companies or required companies, whichever is the smaller. If fully satisfactory records of response are not kept, such credit may be limited to one paid member for each eight claimed to respond. Call or volunteer members sleeping at fire stations may be considered as equivalent to on-duty members in determining night strength, provided they are available for response to alarms for at least nine hours during the night.

Consideration shall be given to periods when the response of call or volunteer members is below average.

Off-shift paid members responding voluntarily on first alarms may be credited on the same basis as call and volunteer members.

Note C. Automatic Aid. The strength of those companies with apparatus credited or creditable as automatic aid under Items 1 and 2 may be considered as the combined least number of full-paid members normally responding with these companies. If these companies are staffed by volunteers, four volunteers may be considered equivalent to one full-paid member, and the strength may be taken as the combined least number of equivalent full-paid members responding with these companies, but the total of these equivalent members and those considered under Note B shall not exceed one-third the required strength of the department receiving aid.

Credit shall depend upon effective operation of the aid plan but shall not reduce the point charge by more than 75 or 90 percent, as may be applicable.

Note D. Off-Shift Response. Credit may be given, on the basis of four off-duty paid members considered equivalent to one on-duty member, for the number of paid members off duty who are required by regulations to respond when called, provided means and arrangements for notification are satisfactory. Past off-shift response experience, including the time taken to return to duty, shall be considered in determining the number of members to be allowed. Credit shall not reduce the point charge remaining after automatic-aid credit has been applied by more than 75 percent.

Note E. Outside Aid. The strength of those companies credited or creditable as outside aid under Items 1 and 2 may be credited as the combined least number of full paid members or equivalent normally responding with these companies. If these companies are manned by call or volunteer members, 4 such members may be considered equivalent to one full-paid member, but the total of these equivalent members and those considered under Notes B and C shall not exceed one-third the required strength of the department receiving aid. Credit shall depend upon effective operation of the aid plan but shall not reduce the point charge remaining after automatic-aid and off-shift response credit has been applied by more than 33 percent.

FIRE RESEARCH AND SAFETY ACT

Public Law 90-259
90th Congress, S. 1124
March 1, 1968

AN ACT

To amend the Organic Act of the National Bureau of Standards to authorize a fire research and safety program, and for other purposes.

Be it enacted by the Senate and House of Representatives of the United States of America in Congress assembled, That this Act may be cited as the "Fire Research and Safety Act of 1968."

TITLE I—FIRE RESEARCH
AND SAFETY PROGRAM

Declaration of Policy

Sec. 101 The Congress finds that a comprehensive fire research and safety program is needed in this country to provide more effective measures of protection against the hazards of death, injury, and damage to property. The Congress finds that it is desirable and necessary for the Federal Government, in carrying out the provisions of this title, to cooperate with and assist public and private agencies. The Congress declares that the purpose of this title is to amend the Act of March 3, 1901, as amended, to provide a national fire research and safety program including the gathering of comprehensive fire data; a comprehensive fire research program; fire safety education and training programs; and demonstrations of new approaches and improvements in fire prevention and control, and reduction of death, personal injury, and property damage. Additionally, it is the sense of Congress that the Secretary should establish a fire research and safety center for administering this title and carrying out its purposes, including appropriate fire safety liaison and coordination.

Authorization of Program

Sec. 102. The Act entitled "An Act to establish the National Bureau of Standards," approved March 3, 1901, as amended (15 U.S.C. 271-278e), is further amended by adding the following sections:

"Sec. 16. The Secretary of Commerce (hereinafter referred to as the 'Secretary') is authorized to—

"(a) Conduct directly or through contracts or grants—

"(1) investigations of fires to determine their causes, frequency of occurrence, severity, and other pertinent factors;

APPENDIX D

"(2)   research into the causes and nature of fires, and the development of improved methods and techniques for fire prevention, fire control, and reduction of death, personal injury, and property damage;

"(3)   educational programs to—

"(A)   inform the public of fire hazards and fire safety techniques and

"(B)   encourage avoidance of such hazards and use of such techniques;

"(4)   fire information reference services, including the collection, analysis, and dissemination of data, research results, and other information, derived from this program or from other sources and related to fire protection, fire control, and reduction of death, personal injury, and property damage;

"(5)   educational and training programs to improve, among other things—

"(A)   the efficiency, operation, and organization of fire services, and

"(B)   the capability of controlling unusual fire-related hazards and fire disasters; and

"(6)   projects demonstrating

"(A)   improved or experimental programs of fire prevention, fire control, and reduction of death, personal injury, and property damage,

"(B)   application of fire safety principles in construction, or

"(C)   improvement of the efficiency, operation, or organization of the fire services.

"(b)   Support by contracts or grants the development, for use by educational and other nonprofit institutions, of—

"(1)   fire safety and fire protection engineering or science curriculums; and

"(2)   fire safety courses, seminars, or other instructional materials and aids for the above curriculums or other appropriate curriculums or courses of instruction.

"Sec. 17.   With respect to the functions authorized by section 16 of this Act—

"(a)   Grants may be made only to States and local governments, other non-Federal public agencies, and nonprofit institutions. Such a grant may be up to 100 per centum of the total cost of the project for which such grant is made. The Secretary shall require, whenever feasible, as a condition of approval of a grant, that the recipient contribute money, facilities, or services to carry out the purpose for which the grant is sought. For the purposes of this section, 'State' means any State of the United States, the

District of Columbia, the Commonwealth of Puerto Rico, the Virgin Islands, Guam, the Canal Zone, American Samoa, and the Trust Territory of the Pacific Islands; and 'public agencies' includes combinations or groups of States or local governments.

"(b)   The Secretary may arrange with and reimburse the heads of other Federal departments and agencies for the performance of any such functions, and as necessary or appropriate, delegate any of his powers under this section or section 16 of this Act with respect to any part thereof, and authorize the redelegation of such powers.

"(c)   The Secretary may perform such functions without regard to section 3648 of the Revised Statutes (31 U.S.C. 529).

"(d)   The Secretary is authorized to request any Federal department or agency to supply such statistics, data, program reports, and other materials as he deems necessary to carry out such functions. Each such department or agency is authorized to cooperate with the Secretary and, to the extent permitted by law, to furnish such materials to the Secretary. The Secretary and the heads of other departments and agencies engaged in administering programs related to fire safety shall, to the maximum extent practicable, cooperate and consult in order to insure fully coordinated efforts.

"(e)   The Secretary is authorized to establish such policies, standards, criteria, and procedures and to prescribe such rules and regulations as he may deem necessary or appropriate to the administration of such functions or this section, including rules and regulations which—

"(1)   provide that a grantee will from time to time, but not less often than annually, submit a report evaluating accomplishments of activities funded under section 16, and

"(2)   provide for fiscal control, sound accounting procedures, and periodic reports to the Secretary regarding the application of funds paid under section 16."

Noninterference with Existing Federal Programs

Sec. 103.   Nothing contained in this title shall be deemed to repeal, supersede, or diminish existing authority or responsibility of any agency or instrumentality of the Federal Government.

Authorization of Appropriations

Sec. 104.   There are authorized to be appropriated, for the purposes of this Act, $5,000,000 for the period ending June 30, 1970.

TITLE II—NATIONAL COMMISSION ON FIRE PREVENTION AND CONTROL

Findings and Purpose

Sec. 201.   The Congress finds and declares that the growing problem of the loss of life and property from fire is a matter of grave national concern;

that this problem is particularly acute in the Nation's urban and suburban areas where an increasing proportion of the population resides but it is also of national concern in smaller communities and rural areas; that as population concentrates, the means for controlling and preventing destructive fires has become progressively more complex and frequently beyond purely local capabilities; and that there is a clear and present need to explore and develop more effective fire control and fire prevention measures throughout the country in the light of existing and foreseeable conditions. It is the purpose of this title to establish a commission to undertake a thorough study and investigation of this problem with a view to the formulation of recommendations whereby the Nation can reduce the destruction of life and property caused by fire in its cities, suburbs, communities, and elsewhere.

Establishment of Commission

Sec. 202.

(a)   There is hereby established the National Commission on Fire Prevention and Control (hereinafter referred to as the "Commission") which shall be composed of twenty members as follows: the Secretary of Commerce, the Secretary of Housing and Urban Development, and eighteen members appointed by the President. The individuals so appointed as members (1) shall be eminently well qualified by training or experience to carry out the functions of the Commission, and (2) shall be selected so as to provide representation of the views of individuals and organizations of all areas of the United States concerned with fire research, safety, control, or prevention, including representatives drawn from Federal, State, and local governments, industry, labor, universities, laboratories, trade associations, and other interested institutions or organizations. Not more than six members of the Commission shall be appointed from the Federal Government. The President shall designate the Chairman and Vice Chairman of the Commission.

(b)   The Commission shall have four advisory members composed of—

>   (1)   two Members of the House of Representatives who shall not be members of the same political party and who shall be appointed by the Speaker of the House of Representatives, and

>   (2)   two members of the Senate who shall not be members of the same political party and who shall be appointed by the President of the Senate.

The advisory members of the Commission shall not participate, except in an advisory capacity, in the formulation of the findings and recommendations of the Commission.

(c)   Any vacancy in the Commission or in its advisory membership shall not affect the powers of the Commission, but shall be filled in the same manner as the original appointment.

Sec. 203.

(a)   The Commission shall undertake a comprehensive study and investigation to determine practicable and effective measures for reducing the destructive effects of fire throughout the country in addition to the steps taken under sections 16 and 17 of the Act of March 3, 1901 (as added by title I of this Act). Such study and investigation shall include, without being limited to—

(1)   a consideration of ways in which fires can be more effectively prevented through technological advances, construction techniques, and improved inspection procedures;

(2)   an analysis of existing programs administered or supported by the departments and agencies of the Federal Government and of ways in which such programs could be strengthened so as to lessen the danger of destructive fires in Government-assisted housing and in the redevelopment of the Nation's cities and communities;

(3)   an evaluation of existing fire suppression methods and of ways for improving the same, including procedures for recruiting and soliciting the necessary personnel;

(4)   an evaluation of present and future needs (including long-term needs) of training and education for fire-service personnel;

(5)   a consideration of the adequacy of current fire communication techniques and suggestions for the standardization and improvement of the apparatus and equipment used in controlling fires;

(6)   an analysis of the administrative problems affecting the efficiency or capabilities of local fire departments or organizations; and

(7)   an assessment of local, State, and Federal responsibilities in the development of practicable and effective solutions for reducing fire losses.

(b)   In carrying out its duties under this section the Commission shall consider the results of the functions carried out by the Secretary of Commerce under sections 16 and 17 of the Act of March 3, 1901 (as added by title I of this Act), and consult regularly with the Secretary in order to coordinate the work of the Commission and the functions carried out under such sections 16 and 17.

(c)   The Commission shall submit to the President and to the Congress a report with respect to its findings and recommendations not later than two years after the Commission has been duly organized.

Powers and Administrative Provisions

Sec. 204.

(a)   The Commission or, on the authorization of the Commission, any subcommittee or member thereof, may, for the purpose of carrying out the provisions of this title, hold hearings, take testimony, and administer oaths or affirmations to witnesses appearing before the Commission or any sub-committee or member thereof.

(*b*) Each department, agency, and instrumentality of the executive branch of the Government, including an independent agency, is authorized to furnish to the Commission, upon request made by the Chairman or Vice Chairman, such information as the Commission deems necessary to carry out its functions under this title.

(*c*) Subject to such rules and regulations as may be adopted by the commission, the Chairman, without regard to the provisions of title 5, United States Code, governing appointments in the competitive service, and without regard to the provisions of chapter 51 and subchapter III of chapter 53 of such title relating to classification and General Schedule pay rates, shall have the power—

(1) to appoint and fix the compensation of such staff personnel as he deems necessary, and

(2) to procure temporary and intermittent services to the same extent as is authorized by section 3109 of title 5, United States Code.

Compensation of Members

Sec. 205.

(*a*) Any member of the Commission, including a member appointed under section 202(b), who is a Member of Congress or in the executive branch of the Government shall serve without compensation in addition to that received in his regular employment, but shall be entitled to reimburse-ment for travel, subsistence, and other necessary expenses incurred by him in connection with the performance of duties vested in the Commission.

(*b*) Members of the Commission, other than those referred to in sub-section (a), shall receive compensation at the rate of $100 per day for each day they are engaged in the performance of their duties as members of the Commission and shall be entitled to reimbursement for travel, subsistence, and other necessary expenses incurred by them in the performance of their duties of the Commission.

Expenses of the Commission

Sec. 206. There are authorized to be appropriated, out of any money in the Treasury not otherwise appropriated, such sums as may be necessary to carry out this title.

Expiration of the Commission

Sec. 207. The Commission shall cease to exist thirty days after the submission of its report under section 203(c).

Recommendations of the National Commission on Fire Prevention and Control

In 1971, President Nixon appointed the eighteen-member Committee of the National Commission on Fire Prevention and Control, established under Title II of the Fire Research and Safety Act. The Secretary of Commerce, Frederick B. Dent, and the Secretary of Housing and Urban Development, James T. Lynn, were the other two members of the Commission. Funding of the Commission also was provided for by Congress in 1971.

- Richard E. Bland, Chairman, Associate Professor, Pennsylvania State University
- W. Howard McClennan, Vice Chairman, President, I. A. of F.F.
- Tommy Arevalo, Lieutenant, El Paso Fire Department
- Percy Bugbee, Honorary Chairman, National Fire Protection Association
- John L. Jablonsky, A.V.P., American Insurance Association
- Albert E. Hole, California State Fire Marshal
- Anne W. Phillips, M.D., Burn Specialist, Massachusetts General Hospital
- Roger M. Freeman, Jr., President, Allendale Mutual Insurance Company
- Ernst R. G. Eckert, Ph.D., University of Minnesota
- Keith E. Klinger, Chief Emeritus, Los Angeles County Fire Department
- Robert A. Hechtman, Ph.D., R. A. Hechtman and Associates
- Louis J. Amabili, Director, Delaware State Fire School
- Peter S. Hackes, National Broadcasting Company, Washington
- William J. Young, Chief, Newington Fire Department, New Hampshire
- Dorothy Duke, Consultant to National Council of Negro Women
- John F. Hurley, Fire Commissioner, Rochester, New York
- John A. Proven, Fire Equipment Manufacturers Association
- Baron Whitaker, President, Underwriters' Laboratories

The Commission held public hearings at selected locations in the country to aid them in their study and investigation. Each hearing discussed a different aspect of the overall problem. Selected witnesses submitted papers and were questioned by the Commission on these various aspects.

APPENDIX E

After two years of public hearings and research, the Commission Report was delivered to President Nixon on May 4, 1973. This twenty-chapter report, *America Burning*, may be purchased from the Superintendent of Documents, U.S. Government Printing Office, Washington. With the exception of Chapter 8, which details statistics on fire losses, and Chapter 20, on what citizens can do, recommendations were presented by the Commission with the goal of reduced fire losses in both life and property.

The recommendations follow:

CHAPTER 1

1. ... the Commission recommends that Congress establish a U.S. Fire Administration to provide a national focus for the Nation's fire problem and to promote a comprehensive program with adequate funding to reduce life and property loss from fire.

2. ... the Commission recommends that a national fire data system be established to provide a continuing review and analysis of the entire fire problem.

CHAPTER 2

3. The Commission recommends that Congress enact legislation to make possible the attainment of 25 burn units and centers and 90 burn programs within the next 10 years.

4. The Commission recommends that Congress, in providing for new burn treatment facilities, make adequate provision for the training and continuing support of the specialists to staff these facilities. Provision should also be made for special training of those who provide emergency care for burn victims in general hospitals.

5. The Commission recommends that the National Institutes of Health greatly augment their sponsorship of research on burns and burn treatment.

6. The Commission recommends that the National Institutes of Health administer and support a systematic program of research concerning smoke inhalation injuries.

CHAPTER 3

7. The Commission recommends that local governments make fire prevention at least equal to suppression in the planning of fire department priorities.

8. The Commission recommends that communities train and utilize women for fire service duties.

9. The Commission recommends that laws which hamper cooperative arrangements among local fire jurisdictions be changed to remove the restrictions.

10. The Commission recommends that every local fire jurisdiction prepare a master plan designed to meet the community's present and future needs in fire protection, to serve as a basis for program budgeting, and to identify and implement the optimum cost-benefit solutions in fire protection.

11. ... the Commission recommends that Federal grants for equipment and training be available only to those fire jurisdictions that operate from a federally approved master plan for fire protection.

12. The Commission recommends that the proposed U.S. Fire Administration act as a coordinator of studies of fire protection methods and assist local jurisdictions in adapting findings to their fire protection planning.

CHAPTER 4

13. The Commission recommends that the proposed U.S. Fire Administration provide grants to local fire jurisdictions for developing master plans for fire protection. Further, the proposed U.S. Fire Administration should provide technical advice and qualified personnel to local fire jurisdictions to help them develop master plans.

CHAPTER 5

14. ... the Commission recommends that the proposed U.S. Fire Administration sponsor research in the following areas: productivity measure of fire departments, job analyses, firefighter injuries, and fire prevention efforts.

15. ... the Commission urges the Federal research agencies, such as the National Science Foundation and the National Bureau of Standards, to sponsor research appropriate to their respective missions within the areas of productivity of fire departments, causes of firefighter injuries, effectiveness of fire prevention efforts, and the skills required to perform various fire department functions.

16. The Commission recommends that the Nation's fire departments recognize advanced and specialized education and hire or promote persons with experience at levels commensurate with their skills.

17. The Commission recommends a program of Federal financial assistance to local fire services to upgrade their training.

18. In the administering of Federal funds for training or other assistance to local fire departments, the Commission recommends that eligibility be limited to those departments that have adopted an effective, affirmative action program related to the employment and promotion of members of minority groups.

19. The Commission recommends that fire departments, lacking emergency ambulance, paramedical, and rescue services consider providing them, especially if they are located in communities where these services are not adequately provided by other agencies.

CHAPTER 6

20. ... the Commission recommends the establishment of a National Fire Academy to provide specialized training in areas important to the fire services and to assist State and local jurisdictions in their training programs.

21. The Commission recommends that the proposed National Fire Academy assume the role of developing, gathering, and disseminating, to State and local arson investigators, information on arson incidents and on advanced methods of arson investigations.

22. The Commission recommends that the National Fire Academy be organized as a division of the proposed U.S. Fire Administration, which would assume responsibility for deciding details of the Academy's structure and administration.

23. The Commission recommends that the full cost of operating the proposed National Fire Academy and subsidizing the attendance of fire service members be borne by the Federal Government.

CHAPTER 7

24. The Commission urges the National Science Foundation, in its Experimental Research and Development Incentives Program, and the National Bureau of Standards, in its Experimental Technology Incentives Program, to give high priority to the needs of the fire services.

25. The Commission recommends that the proposed U.S. Fire Administration review current practices in terminology, symbols, and equipment descriptions, and seek to introduce standardization where it is lacking.

26. The Commission urges rapid implementation of a program to improve breathing apparatus systems and expansion of the program's scope where appropriate.

27. The Commission recommends that the proposed U.S. Fire Administration undertake a continuing study of equipment needs of the fire services, monitor research and development in progress, encourage needed research and development, disseminate results, and provide grants to fire departments for equipment procurement to stimulate innovation in equipment design.

28. ... the Commission urges the Joint Council of National Fire Service Organizations to sponsor a study to identify shortcomings of firefighting equipment and the kinds of research, development, or technology transfer that can overcome the deficiencies.

CHAPTER 8

No recommendations.

CHAPTER 9

29. The Commission recommends that research in the basic processes of ignition and combustion be strongly increased to provide a foundation for developing improved test methods.

30. The Commission recommends that the new Consumer Product Safety Commission give a high priority to the combustion hazards of materials in their end use.

31. ... the Commission recommends that the present fuel load study sponsored by the General Services Administration and conducted by the

National Bureau of Standards be expanded to update the technical study of occupancy fire loads.

32. The Commission recommends that flammability standards for fabrics be given high priority by the Consumer Product Safety Commission.

33. The Commission recommends that all States adopt the Model State Fireworks Law of the National Fire Protection Association, thus prohibiting all fireworks except those for public displays.

34. The Commission recommends that the Department of Commerce be funded to provide grants for studies of combustion dynamics and the means of its control.

35. The Commission recommends that the National Bureau of Standards and the National Institutes of Health cooperatively devise and implement a set of research objectives designed to provide combustion standards for materials to protect human life.

CHAPTER 10

36. The Commission urges the National Bureau of Standards to assess current progress in fire research and define the areas in need of additional investigation. Further, the Bureau should recommend a program for translating research results into a systematic body of engineering principles and, ultimately, into guidelines useful to code writers and building designers.

37. The Commission recommends that the National Bureau of Standards, in cooperation with the National Fire Protection Association and other appropriate organizations, support research to develop guidelines for a systems approach to fire safety in all types of buildings.

38. ... the Commission recommends that, in all construction involving Federal money, awarding of those funds be contingent upon the approval of a fire safety systems analysis and a fire safety effectiveness statement.

39. The Commission urges the Consumer Product Safety Commission to give high priority to matches, cigarettes, heating appliances, and other consumer products that are significant sources of burn injuries, particularly products for which industry standards fail to give adequate protection.

40. The Commission recommends to schools giving degrees in architecture and engineering that they include in their curricula at least one course in fire safety. Further, we urge the American Institute of Architects, professional engineering societies, and State registration boards to implement this recommendation.

41. The Commission urges the Society of Fire Protection Engineers to draft model courses for architects and engineers in the field of fire protection engineering.

42. The Commission recommends that the proposed National Fire Academy develop short courses to educate practicing designers in the basics of fire safety design.

CHAPTER 11

43. The commission recommends that all local governmental units in the United States have in force an adequate building code and fire prevention code or adopt whichever they lack.

44. The Commission recommends that local governments provide the competent personnel, training programs for inspectors, and coordination among the various departments involved to enforce effectively the local building and fire prevention codes. Representatives from the fire department should participate in reviewing the fire safety aspects of plans for new building construction and alterations to old buildings.

45. The Commission recommends that, as the model code of the International Conference of Building Officials has already done, all model codes specify at least a single-station early-warning detector oriented to protect sleeping areas in every dwelling unit. Further, the model codes should specify automatic fire extinguishing systems and early-warning detectors for high-rise buildings and for low-rise buildings in which many people congregate.

CHAPTER 12

46. The Commission recommends that the National Transportation Safety Board expand its efforts in issuance of reports on transportation accidents so that the information can be used to improve transportation fire safety.

47. The Commission recommends that the Department of Transportation work with interested parties to develop a marking system, to be adopted nationwide, for the purpose of identifying transportation hazards.

48. The Commission recommends that the proposed National Fire Academy disseminate to every fire jurisdiction appropriate educational materials on the problems of transporting hazardous materials.

49. The Commission recommends the extension of the Chem-Trec system to provide ready access by all fire departments and to include hazard control tactics.

50. ... the Commission recommends that the Department of the Treasury establish adequate fire regulations, suitably enforced, for the transportation, storage, and transfer of hazardous materials in international commerce.

51. The Commission recommends that the Department of Transportation set mandatory standards that will provide fire safety in private automobiles.

52. The Commission recommends that airport authorities review their fire-fighting capabilities and, where necessary, formulate appropriate capital improvement budgets to meet current recommended aircraft rescue and fire-fighting practices.

53. The Commission recommends that the Department of Transportation undertake a detailed review of the Coast Guard's responsibilities, authority, and standards relating to marine fire safety.

54. The Commission recommends that the railroads begin a concerted effort to reduce rail-caused fires along the Nation's rail system.

55. ... the Commission recommends that the Urban Mass Transportation Administration require explicit fire safety plans as a condition for all grants for rapid transit systems.

CHAPTER 13

56. ... the Commission recommends that rural dwellers and others living at a distance from fire departments install early-warning detectors and alarms to protect sleeping areas.

57. The Commission recommends that U.S. Department of Agriculture assistance to (community fire protection facilities) projects be contingent upon an approved master plan for fire protection for local fire jurisdictions.

CHAPTER 14

58. ... the Commission recommends that the proposed U.S. Fire Administration join with the Forest Service, U.S.D.A., in exploring means to make fire safety education for forest and grassland protection more effective.

59. The Commission recommends that the Council of State Governments undertake to develop model State laws relating to fire protection in forests and grasslands.

60. The Commission urges interested citizens and conservation groups to examine fire laws and their enforcement in their respective States and to press for strict compliance.

61. The Commission recommends that the Forest Service, U.S.D.A., develop the methodology to make possible nationwide forecasting of fuel buildup as a guide to priorities in wildland management.

62. The Commission supports the development of a National Fire Weather Service in the National Oceanic and Atmospheric Administration and urges its acceleration.

CHAPTER 15

63. The Commission recommends that the Department of Health, Education, and Welfare include in accreditation standards fire safety education in the schools throughout the school year. Only schools presenting an effective fire safety education program should be eligible for any Federal financial assistance.

64. The Commission recommends that the proposed U.S. Fire Administration sponsor fire safety education courses for educators to provide a teaching cadre for fire safety education.

65. The Commission recommends to the States the inclusion of fire safety education in programs educating future teachers and the requirement of knowledge of fire safety as a prerequisite for teaching certification.

66. The Commission recommends that the proposed U.S. Fire Administration develop a program, with adequate funding, to assist, augment, and evaluate existing public and private fire safety education efforts.

67. ... the Commission recommends that the proposed U.S. Fire Administration, in conjunction with the Advertising Council and the National Fire Protection Association, sponsor an all-media campaign of public service advertising designed to promote public awareness of fire safety.

68. The Commission recommends that the proposed U.S. Fire Administration develop packets of educational materials appropriate to each occupational category that has special needs or opportunities in promoting fire safety.

CHAPTER 16

69. The Commission supports the Operation EDITH (Exit Drills In The Home) plan and recommends its acceptance and implementation both individually and community-wide.

70. The Commission recommends that annual home inspections be undertaken by every fire department in the Nation. Further, Federal financial assistance to fire jurisdictions should be contingent upon their implementation of effective home fire inspection programs.

71. The Commission urges Americans to protect themselves and their families by installing approved early-warning fire detectors and alarms in their homes.

72. ... the Commission recommends that the insurance industry develop incentives for policyholders to install approved early-warning fire detectors in their residences.

73. The Commission urges Congress to consider amending the Internal Revenue Code to permit reasonable deductions from income tax for the cost of installing approved detection and alarm systems in homes.

74. ... the Commission recommends that the proposed U.S. Fire Administration monitor the progress of research and development on early-warning detection systems in both industry and government and provide additional support for research and development where it is needed.

75. The Commission recommends that the proposed U.S. Fire Administration support the development of the necessary technology for improved automatic extinguishing systems that would find ready acceptance by Americans in all kinds of dwelling units.

76. The Commission recommends that the National Fire Protection Association and the American National Standards Institute jointly review the Standard for Mobile Homes and seek to strengthen it, particularly in such areas as interior finish materials and fire detection.

77. The Commission recommends that all political jurisdictions require compliance with the NFPA/ANSI standard for mobile homes together with additional requirements for early-warning fire detectors and improved fire resistance of materials.

78. The Commission recommends that State and local jurisdictions adopt the NFPA Standard on Mobile Home Parks as a minimum mode of protection for the residents of these parks.

CHAPTER 17

79. The Commission strongly endorses the provisions of the Life Safety Code which require specific construction features, exit facilities, and fire detection systems in child day care centers and recommends that they be adopted and enforced immediately by all the States as a minimum requirement for licensing of such facilities.

80. The Commission recommends that early-warning detectors and total automatic sprinkler protection or other suitable automatic extinguishing systems be required in all facilities for the care and housing of the elderly.

81. The Commission recommends to Federal agencies and the States that they establish mechanisms for annual review and rapid upgrading of their fire safety requirements for facilities for the aged and infirm, to a level no less stringent than the current NFPA Life Safety Code.

82. The Commission recommends that the special needs of the physically handicapped and elderly in institutions, special housing, and public buildings be incorporated into all fire safety standards and codes.

83. The Commission recommends that the States provide for periodic inspection of facilities for the aged and infirm, either by the State's fire marshal's office or by local fire departments, and also require approval of plans for new facilities and inspection by a designated authority during and after construction.

84. The Commission recommends that The National Bureau of Standards develop standards for the flammability of fabric materials commonly used in nursing homes with a view to providing the highest level of fire resistance compatible with the state-of-the-art and reasonable costs.

85. The Commission recommends that political subdivisions regulate the location of nursing homes and housing for the elderly and require that fire alarm systems be tied directly and automatically to the local fire department.

CHAPTER 18

86. The Commission recommends that the Federal Government retain and strengthen its programs of fire research for which no non-governmental alternatives exist.

87. ... the Commission recommends that the Federal budget for research connected with fire be increased by $26 million.

88. ... the Commission recommends that associations of material and product manufacturers encourage their member companies to sponsor research directed toward improving the fire safety of the built environment.

CHAPTER 19

89. ... the Commission recommends that the proposed U.S. Fire Administration be located in the Department of Housing and Urban Development.

90. The Commission recommends that Federal assistance in support of State and local fire service programs be limited to those jurisdictions complying with the National Fire Data System reporting requirements.

CHAPTER 20

No recommendations.

INDEX

Manpower, 78–79, 318–320
Masks and respiratory equipment, 48, 199–202
Materials (*see* Combustible materials; Hazardous materials; Radioactive materials)
Matter (substance), 83–85
Mechanical air foam, 167
  illustrated, 166
Mechanical heat energy, 108–109
Metals, 92–93, 257
Mill construction (heavy timber), 143–144
Molecules, 83–84, 86
Multiple-death fires, 25
  table, 26
Multi-purpose dry chemical extinguisher, 169

National Board of Fire Underwriters, 55
National Commission on Fire Prevention and Control, 48, 323–326, 327–328
  recommendations of, 331–339
National Fire Protection Association (NFPA), 54–56
  fire-apparatus standards, 178, 185, 188, 190, 199
  fire-incident reporting, 21–23
  hazardous material guidelines, 243, 246–248, 251
    illustrated, 247, 248
  national statistics on fire loss, 21, 23–24
    tables, 26–28
  occupancy classification, 151
  standard types of building construction, 137
Neutrons, 87
Night watch, 1
Nitrogen, 92, 95–96
Noncombustible construction, 143–144
Nozzles, 188, 229
  nozzle pressure, 229
Nuclear heat energy, 109

Objectives of fire department, 65, 207
Occupancy:
  classification of fires by, 30–31
    table, 32–35
  content-hazard classification, 152
  large-loss fires, table, 36–38
  safety of life, code classification, 151–152

Ordinary construction, 144–145
Organic chemical compounds, 159
Organizational charts, 68–70
Overhaul, 221
Oxidation process, 96–97
Oxidizing materials, 97, 160, 164, 171, 257
Oxygen, 92, 93, 95–98

Panic, 215–216
Parapet walls, 149
  illustrated, 150
Plant fire brigades, 278
Platforms, elevating, 15, 192–193
Pompier ladder, 215n.
Portable fire extinguishers, 173, 278–279
Positioning units at a fire, 232–238
  illustrated, 237, 239
Potassium, 93, 171, 257
Potassium bicarbonate, 163, 169
Preaction sprinkler systems, 268
Preconnected hose lines, 187
Preplan inspection, 31, 71, 215, 288–291
Privately owned fire departments, 53
Procedural outline, 208–223
  calling for help, 213–215
  covering of exposures and confinement, 216–218
  extinguishment, 220–221
  overhauling, 221–222
  salvage, 222–223
  saving of life, 215–216
  size-up of situation, 210–213
  ventilation and forcible entry, 218–220
Products of combustion, 121, 129
Propane, 254–255
Proprietary alarm and detection system, 274
Protective clothing, 203
Protons, 87
Public relations:
  as an administrative function, 71–72
  in preplanning inspections, 289
Pumpers (*see* Fire engines)
Pumps:
  capacities of, 15, 178
  centrifugal, 180–184
  positive-displacement, 179
Pyrolysis, 117–118

Radiant heat, 111–112
  illustrated, 114
Radiation rays, 258–259
Radioactive materials, 93, 258–259